Theoretical Concepts of Quantum Mechanics

Editor

Renard Nowak

scitus
academics

Theoretical Concepts of Quantum Mechanics
Edited by **Renard Nowak**

ISBN: 978-1-68117-533-1
Library of Congress Control Number: 2016942933

SCITUS Academics LLC,
www.scitusacademics.com
Box No. 4766, 616 Corporate Way,
Suite 2, Valley Cottage,
NY 10989

Notice

Printed in United States of America on Acid Free Paper

THEORETICAL CONCEPTS OF QUANTUM MECHANICS

PREFACE

Quantum theory as a medical revolution profoundly influenced human thinking about the universe and the forces of nature. Quantum mechanics is the very small physics department. This results in some really weird conclusions about the real physical world. Many of the formulas of classical mechanics, which describe exactly how things move at everyday sizes and speeds, are no longer helpful on the scale of electrons and atoms. Quantum mechanics (QM) developed over many decades, starting as a pair of debatable mathematical explanations of experiments that could not be explained by the math of classical mechanics. It began at the turn of the twentieth century; at exactly the same time that Albert Einstein published his theory of relativity, his own mathematical revolution in physics, which speaks of high-speed motion. Quantum mechanics attempts to explain and take into account the properties of atoms and molecules and their constituents-electrons, neutrons, protons, and more esoteric particles such as gluons and quarks. These properties include the interactions of the particles with each other and with electromagnetic radiation. An important feature of quantum mechanics is that it is usually impossible to determine a system without disturbing it, even in principle; the detailed nature of this disturbance and the exact point at which it occurs are controversial and obscure.

Representing a rich explanation of the introduction, scientific history of quantum mechanics and relativistic quantum mechanics, and theory of fields and various methods for solving the Schrodinger equation; this book will serve as valuable tool for students and researchers.

TABLE OF CONTENTS

7

A NEW FORMULATION OF QUANTUM MECHANICS

A. I. Arbab[1], Faisal A. Yassein[2]

[1]Department of Physics, Faculty of Science, University of Khartoum, Khartoum, Sudan

[2]Department of Physics, Faculty of Science, Alneelain University, Khartoum, Sudan

ABSTRACT

A new formulation of quantum mechanics based on differential commutator brackets is developed. We have found a wave equation representing the fermionic particle. In this formalism, the continuity equation mixes the Klein-Gordon and Schrodinger probability density while keeping the Klein-Gordon and Schrodinger current unaltered. We have found time and space transformations under which Dirac's equation is invariant. The invariance of Maxwell's equations under these transformations shows that the electric and magnetic fields of a moving charged particle are perpendicular to the velocity of the propagating particle. This formulation agrees with the quaternionic formulation recently developed by Arbab.

KEYWORDS

Mathematical Formulation; Quantum Mechanics; Differential Commutator Brackets

1. INTRODUCTION

Schrodinger's equation was used to explain and describe all phenomena in atomic physics. However, after the development of the theory of special relativity by Einstein, there was a need to unify quantum mechanics and special relativity into a single Relativistic Quantum Theory. Despite the success of Schrodinger's equation in describing quite accurately the Hydrogen spectrum and giving correct predictions for a large amount of spectral data, this equation is not invariant under Lorenz transformations. In other words Schrodinger's equation is not relativistic and is only an approximation valid at the non-relativistic limit when the velocities of the particles involved are much smaller than the speed of light.

Quantum mechanics has been formulated by assigning an operator for any dynamical observable. In Heisenberg formalism, the operator is governed by a commutator bracket. The fundamental commutator bracket relates to the position, and momentum is given by $[x, p_x] = i\hbar$. The commutator bracket generalizes the Poisson bracket of classical mechanics. If an operator commutes with Hamiltonian of the system, then the dynamical variable corresponding to that operator is said to be conserved. An equation compatible with Lorentz transformation guarantees its applicability to any inertial frame. Such an equation is symmetric in space-time. Thus a symmetric space-time formulation of any theory will generally guarantee the universality of the theory. However, Schrodinger equations doesn't exhibit this feature because it is not symmetric in space and time. To remedy this problem, Klein and Gordon looked for an equation which is second order in space and time and consequently obtained the Klein-Gordon equation (KG). The probability density in this theory is found to be non-positive definite. Consequently, Dirac thought for a linear equation in space and time that has no such a problem. He obtained the familiar Dirac equation with a positive definite probability. However, the probability in KG formalism is later on (from a theoretic field point of view) interpreted as a charge density rather than a probability density which could be positive or negative [1-3].

With these motivation, we adopt a differential commutator bracket involving first order space and time derivative operators to formulate the Maxwell equations and quantum mechanics. This is in addition to our recent quarternionic formulation of physical laws, where we have shown that many physical equations are found to emerge from a unified form of physical variables [4]. Moreover, using quaternions, we have recently shown that quantum mechanics can be formulated in a set of three equations [5]. In such a formulation, the Dirac and Klein-Gordon equations emerge from a set of three equations obtained from the application of an eigen-value problem of the linear momentum.

We aim in this paper to derive the equation of motion of the quantum system by applying the vanishing differential commutator brackets. It is interesting to note that these commutator brackets are Lorentz invariant.

Moreover, $[\Delta t,\Delta x]=[\Delta t',\Delta x']$, where t is time, $\nabla=\frac{\partial}{\partial x}\hat{i}+\frac{\partial}{\partial y}\hat{j}+\frac{\partial}{\partial z}\hat{k}$, $\Delta x=x_2-x_1$ and t',x' are the moving time and space coordinates. We know that the second order partial derivatives commute for space-space variables. We don't assume here that this property is a priori for space and time. To guarantee this, we eliminate the time derivative of a quantity that is acted by a space (∇) derivative followed by a time derivative, and vice versa. In expanding the differential commutator brackets, we don't commute time and space derivative, but rather eliminate the time derivative by the space derivative, and vice versa. These linear differential commutator brackets may enlighten us to quantize these physical quantities. By employing the differential commutator brackets of the vector A and scalar potential ϕ, we have derived Maxwell equations without invoking any a priori physical law [6]. We would like here to apply the differential commutator brackets to explore quantum mechanics.

2. DIFFERENTIAL COMMUTATORS ALGEBRA

Define the three linear differential commutator brackets as follows [6];

$$\left[\frac{\partial}{\partial t},\nabla\right]=0,\quad \left[\frac{\partial}{\partial t},\nabla\cdot\right]=0,\quad \left[\frac{\partial}{\partial t},\nabla\times\right]=0, \tag{1}$$

where ∇ and $\frac{\partial}{\partial t}$ are the space and time derivatives.

For a scalar ϕ and a vector A, one finds that:

$$\left[\frac{\partial}{\partial t},\nabla\right]\phi=\frac{\partial}{\partial t}(\nabla\phi)-\nabla\left(\frac{\partial\phi}{\partial t}\right), \tag{2}$$

And

$$\left[\frac{\partial}{\partial t},\nabla\cdot\right]\boldsymbol{A}=\frac{\partial}{\partial t}(\nabla\cdot\boldsymbol{A})-\nabla\cdot\left(\frac{\partial\boldsymbol{A}}{\partial t}\right),$$
$$\left[\frac{\partial}{\partial t},\nabla\times\right]\boldsymbol{A}=\frac{\partial}{\partial t}(\nabla\times\boldsymbol{A})-\nabla\times\left(\frac{\partial\boldsymbol{A}}{\partial t}\right). \tag{3}$$

Moreover, one can show that:

$$\left[\frac{\partial}{\partial t},\nabla\cdot\right](\phi\boldsymbol{A})=\phi\left\{\left[\frac{\partial}{\partial t},\nabla\cdot\right]\boldsymbol{A}\right\}+\left\{\left[\frac{\partial}{\partial t},\nabla\right]\phi\right\}\cdot\boldsymbol{A},$$
$$\left[\frac{\partial}{\partial t},\nabla\times\right](\phi\boldsymbol{A})=\phi\left\{\left[\frac{\partial}{\partial t},\nabla\times\right]\boldsymbol{A}\right\}+\left\{\left[\frac{\partial}{\partial t},\nabla\right]\phi\right\}\times\boldsymbol{A},$$
$$\left[\frac{\partial}{\partial t},\nabla\cdot\right](\boldsymbol{A}\times\boldsymbol{B})=\boldsymbol{B}\cdot\left\{\left[\frac{\partial}{\partial t},\nabla\times\right]\boldsymbol{A}\right\}-\boldsymbol{A}\cdot\left\{\left[\frac{\partial}{\partial t},\nabla\times\right]\boldsymbol{B}\right\}. \tag{4}$$

The differential commutator bracket satisfies the distribution rule:

$$\left[\hat{A}\hat{B},\hat{C}\right]=\hat{A}\left[\hat{B},\hat{C}\right]+\left[\hat{A},\hat{C}\right]\hat{B}, \tag{5}$$

where $\hat{A},\hat{B},\hat{C}$ stand for either ∇ or $\frac{\partial}{\partial t}$.

It is evident that the differential commutator brackets identities follow the same ordinary vector identities. We call the three differential commutator brackets in Equation (1) the grad-commutator bracket, the dotcommutator bracket, and the cross-commutator bracket, respecttively. The prime idea here is to replace the time derivative of a quantity by the space derivative ∇ of another quantity, and vice versa, so that the time derivative of a quantity is followed by a time derivative with which it commutes. We assume here that space and time derivatives don't commute. With this minimal assumption, we have shown here that all physical laws are determined by vanishing differential commutator bracket.

3. THE CONTINUITY EQUATION

Using quaternionic algebra [7-9], we have recently found that generalized continuity equations can be written as [5].

$$\nabla \cdot \boldsymbol{J} + \frac{\partial \rho}{\partial t} = 0, \tag{6}$$

$$\nabla\left(\rho c^2\right) + \frac{\partial \boldsymbol{J}}{\partial t} = 0, \tag{7}$$

And

$$\nabla \times \boldsymbol{J} = 0, \tag{8}$$

where c, ρ and $\boldsymbol{J}$ are the speed of light, current density, and probability density, respectively. Now consider the dot-commutator bracket of $\rho \boldsymbol{J}$.

$$\left[\frac{\partial}{\partial t}, \nabla \cdot\right](\rho \boldsymbol{J}) = \frac{\partial}{\partial t}(\nabla \cdot (\rho \boldsymbol{J})) - \nabla \cdot \left(\frac{\partial (\rho \boldsymbol{J})}{\partial t}\right) = 0. \tag{9}$$

Using Equations (6-8), and the vector identities

$$\begin{aligned} &\nabla \cdot (\phi \boldsymbol{G}) = (\nabla \phi) \cdot \boldsymbol{G} + \phi (\nabla \cdot \boldsymbol{G}), \\ &\nabla \times (\nabla \times \boldsymbol{G}) = \nabla (\nabla \cdot \boldsymbol{G}) - \nabla^2 \boldsymbol{G} \end{aligned} \tag{10}$$

one obtains

$$\left[\frac{\partial}{\partial t},\nabla\cdot\right](\rho\boldsymbol{J})$$
$$=c^2\rho\left(\frac{1}{c^2}\frac{\partial^2\rho}{\partial t^2}-\nabla^2\rho\right)+\left(\frac{1}{c^2}\frac{\partial\boldsymbol{J}}{\partial t^2}-\nabla^2\boldsymbol{J}\right)\cdot\boldsymbol{J}=0. \tag{11}$$

For arbitrary ρ and $\boldsymbol{J}$, Equation (11) yields the two wave equations

$$\frac{1}{c^2}\frac{\partial^2\rho}{\partial t^2}-\nabla^2\rho=0, \tag{12}$$

And

$$\frac{1}{c^2}\frac{\partial\boldsymbol{J}}{\partial t^2}-\nabla^2\boldsymbol{J}=0. \tag{13}$$

Equations (12) and (13) are also obtained utilizing the quaternionic formulation following reference [5]. Hence, the wave equations of ρ and $\boldsymbol{J}$, in our present brackets formulation, are equivalent to

$$\left[\frac{\partial}{\partial t},\nabla\cdot\right](\rho\boldsymbol{J})=0. \tag{14}$$

Equations (12) and (13) show that the charge and current densities satisfy a wave traveling at the speed of light in vacuum. It is remarkable to know that these two equations are already obtained in reference [5].

4. QUANTUM MECHANICS

Consider a particle described by the four vector $\boldsymbol{\Psi}=\left(\frac{i}{c}\psi_0,\boldsymbol{\psi}\right)$. This is equivalent to spinor representation of ordinary quantum mechanics. We have recently developed a quaternionic quantum mechanics dealing with such a four vector [7-9]. The evolution of this four vector is given by the three equations [7-9]

$$\nabla\cdot\boldsymbol{\psi}-\frac{1}{c^2}\frac{\partial\psi_0}{\partial t}-\frac{m_0}{\hbar}\psi_0=0, \tag{15}$$

$$\nabla\psi_0-\frac{\partial\boldsymbol{\psi}}{\partial t}-\frac{m_0c^2}{\hbar}\boldsymbol{\psi}=0, \tag{16}$$

And

$$\nabla\times\boldsymbol{\psi}=0, \tag{17}$$

where m_0 and $\hbar$ are the quasi-particle mass and Planck constant, respectively. Equations (15-17) yield the two wave equations [5].

$$\frac{1}{c^2}\frac{\partial^2\boldsymbol{\psi}}{\partial t^2}-\nabla^2\boldsymbol{\psi}+2\left(\frac{m_0}{\hbar}\right)\frac{\partial\boldsymbol{\psi}}{\partial t}+\left(\frac{m_0c}{\hbar}\right)^2\boldsymbol{\psi}=0, \tag{18}$$

and

$$\frac{1}{c^2}\frac{\partial^2\psi_0}{\partial t^2}-\nabla^2\psi_0+2\left(\frac{m_0}{\hbar}\right)\frac{\partial\psi_0}{\partial t}+\left(\frac{m_0c}{\hbar}\right)^2\psi_0=0. \tag{19}$$

Using the transformation,

$$\frac{\partial}{\partial\tau}=\frac{\partial}{\partial t}+\frac{m_0c^2}{\hbar}, \tag{20}$$

so that Equations (15) and (16) become,

$$\nabla\cdot\boldsymbol{\psi}=\frac{1}{c^2}\frac{\partial\psi_0}{\partial\tau},\quad \nabla\psi_0=\frac{\partial\boldsymbol{\psi}}{\partial\tau}. \tag{21}$$

Employing Equation (20), Equations (18) and (19) are transformed into the wave equations

$$\Pi^2\psi_0 = 0, \ \ \Pi^2\psi = 0, \ \Pi^2 = \frac{1}{c^2}\frac{\partial^2}{\partial\tau^2} - \nabla^2. \tag{22}$$

Equations (18) and (19) can be obtained from the Einstein's energy equation by setting $E' = E + im_0c^2$, where $E'^2 = p^2c^2$ and using the familiar quantum mechanical operator replacements, viz., $\hat{p} = -i\hbar\nabla$ and $\hat{E} = i\hbar\frac{\partial}{\partial t}$. $E' = cp$ is an equation for a massless particle. This is also evident from Equation (22). Thus, it is interesting that a massive particle can be transformed into a massless particle using Equation (20). Since energy is a real quantity, this equation is physically acceptable if it describes a particle with imaginary mass. In this case the energy equations split into two parts; one with $E' = E + m_0c^2$ and the other with energy $E' = E - m_0c^2$. Such energies can describe the state of a particle and antiparticle. A hypotheticcal particle with an imaginary mass moving at a speed higher than the speed of light in vacuum is known as tachyon [10]. Hence, our above equation can be used to treat the motion of tachyons. This implies that our equations, Equations (24) and (27) can be applied to tachyons. Some scientists propose that neutrino can be a tachyonic fermion [11]. We know that the Cherenkov radiation is emitted from a particle moving in a medium with a speed larger than the speed of light in vacuum. When the speed exceeds the speed of light in a vacuum, the extra energy acquired by the particle is transformed in radiation. This can happen momentarily for a particle keeping its total energy conserved. Thus, the excess energy (speed) is such that it compensates the dissipations.

Now consider the cross-commutator bracket of $\psi_0\psi$,

$$\left[\frac{\partial}{\partial t}, \nabla\times\right](\psi_0\psi) = \frac{\partial}{\partial t}(\nabla\times(\psi_0\psi)) - \nabla\times\left(\frac{\partial(\psi_0\psi)}{\partial t}\right) = 0. \tag{23}$$

Using Equations (15), (16), and (17), and the vector identities,

$$\nabla\times(\phi \boldsymbol{G}) = (\nabla\phi)\times\boldsymbol{G} + \phi(\nabla\times\boldsymbol{G}), \tag{24}$$

yield the wave equation,

$$\frac{1}{c^2}\frac{\partial^2\psi}{\partial t^2} - \nabla^2\psi + 2\left(\frac{m_0}{\hbar}\right)\frac{\partial\psi}{\partial t} + \left(\frac{m_0 c}{\hbar}\right)^2\psi = 0. \tag{25}$$

Similarly, the dot-commutator bracket of $\psi_0\psi$,

$$\left[\frac{\partial}{\partial t}, \nabla\cdot\right](\psi_0\psi) = \frac{\partial}{\partial t}(\nabla\cdot(\psi_0\psi)) - \nabla\cdot\left(\frac{\partial(\psi_0\psi)}{\partial t}\right) = 0. \tag{26}$$

Upon using Equations (10), (15), and (16), one obtains the wave equation of ψ_0,

$$\frac{1}{c^2}\frac{\partial^2\psi_0}{\partial t^2} - \nabla^2\psi_0 + 2\left(\frac{m_0}{\hbar}\right)\frac{\partial\psi_0}{\partial t} + \left(\frac{m_0 c}{\hbar}\right)^2\psi_0 = 0. \tag{27}$$

It is interesting to see that Equations (25), and (27) are the same as Equations (18), and (19) obtained from quaternionic manipulation. We thus write Equations (25), and (27) as,

$$\left[\frac{\partial}{\partial t}, \nabla\times\right](\psi_0\psi) = 0, \quad \left[\frac{\partial}{\partial t}, \nabla\cdot\right](\psi_0\psi) = 0. \tag{28}$$

5. DIRAC'S EQUATION

Dirac's equation can be written in the form [1-3],

$$\frac{1}{c}\frac{\partial\psi}{\partial t} + \boldsymbol{\alpha}\cdot\nabla\psi + \frac{i m_0 c\beta}{\hbar}\psi = 0. \tag{29}$$

Consider the differential commutator bracket

$$\left[\frac{\partial}{\partial t},\boldsymbol{\alpha}\cdot\nabla\right]\psi=\frac{\partial}{\partial t}(\boldsymbol{\alpha}\cdot\nabla\psi)-(\boldsymbol{\alpha}\cdot\nabla)\frac{\partial\psi}{\partial t}=0. \tag{30}$$

Using Equation (29), Equation (30) yields,

$$\frac{1}{c^2}\frac{\partial^2\psi}{\partial t^2}-\nabla^2\psi+2\left(\frac{m_0 i\beta}{\hbar}\right)\frac{\partial\psi}{\partial t}-\left(\frac{m_0 c}{\hbar}\right)^2\psi=0, \tag{31}$$

where we have used the fact that $\beta=\begin{pmatrix}1&0\\0&-1\end{pmatrix}$, $\alpha^2=\beta^2=1$ $\alpha=\begin{pmatrix}0&\sigma\\\sigma&0\end{pmatrix}$, and $\boldsymbol{\sigma}$ are the Pauli matrices. Equation (31) can be obtained from Equation (29) by squaring it. This equation can be compared with the KleinGordon equation of spin-0 particles

$$\frac{1}{c^2}\frac{\partial^2\psi}{\partial t^2}-\nabla^2\psi+\left(\frac{m_0 c}{\hbar}\right)^2\psi=0.$$

Equation (31) is another form of Dirac's equation exhibiting the wave nature of spin-1/2 particles explicitly. Using the transformation,

$$\frac{\partial}{\partial\eta}=\frac{\partial}{\partial t}+i\frac{m_0c^2}{\hbar}\beta, \tag{32}$$

Equation (31) can be written as,

$$\frac{1}{c^2}\frac{\partial^2\psi}{\partial\eta^2}-\nabla^2\psi=0. \tag{33}$$

This is a wave equation for a massless particle. Thus, a particle annihilates (loses its mass) after a time interval of $\Delta t=\frac{\hbar}{m_0c^2}$ and then created (acquired a mass). It is interesting to notice that during such a period of time, energy can be violated as endorsed by the Heisenberg's uncertainty relation ($\Delta t\Delta E\geq\hbar$)where $\Delta E=2m_0c^2$. This also applies to the particles as defined by Equation (20). Equation (31) describes the behavior of a particle of a definite mass m_0. After a time of $\Delta t = \frac{\hbar}{m_0 c^2}$ the particle becomes a wave with energy $E' = pc$ governed by Equation (33). The particle interacts with the vacuum in such a way that when the particle becomes a wave (annihilates) gives its mass energy to the vacuum, and restores it after a time of Δt as defined before becoming a particle once again. This is the essence of the oscillatory motion as known as zitterbewegung motion [1-3]. This result supports the fact that there is a vacuum fluctuation associated with the particle. This means when a particle becomes a wave it gives its mass to the vacuum and restores it when becomes a corpuscule. Thus, the corpuscular and wave nature (duality) of a particle is concomitant with the particle motion. Since ψ is a four components spinor, we can write it in terms of two components doublets, viz., $\psi = \begin{pmatrix} \psi_+ \\ \psi_- \end{pmatrix}$. Substituting these decomposed spinors in Equation (31), one obtains the two equations,

$$\frac{1}{c^2}\frac{\partial^2 \psi_+}{\partial t^2} - \nabla^2 \psi_+ + 2\left(\frac{m_0 i}{\hbar}\right)\frac{\partial \psi_+}{\partial t} - \left(\frac{m_0 c}{\hbar}\right)^2 \psi_+ = 0, \tag{34}$$

And

$$\frac{1}{c^2}\frac{\partial^2 \psi_-}{\partial t^2} - \nabla^2 \psi_- - 2\left(\frac{m_0 i}{\hbar}\right)\frac{\partial \psi_-}{\partial t} - \left(\frac{m_0 c}{\hbar}\right)^2 \psi_- = 0, \tag{35}$$

Equations (34) and (35) imply two energy solutions, one with $E' = E - m_0 c^2$ and the other with energy $E' = E + m_0 c^2$. This is also evident from using the Einstein energy-momentum equation ($E^2 = p^2 c^2 + m_0^2 c^4$). The two energy states may define a particle and an antiparticle. Since the time factor in the wavefunction is of the form $\exp(-iEt/\hbar)$, the new wavefunction with the new time (η) will become $\exp(-iE'\eta/\hbar)$, where

$$\eta = t - i\frac{m_0 c^2}{\hbar}, \tag{36}$$

is a complex time, as evident from Equation (33). It can be seen as a rotation of the real time by a phase into a complex plane. Such an effect arises from the very nature of the particle when propagating in space-time. The third term in Equation (31) represents a dissipation that may result from the motion of the particle in space (ether). Hence, any massive particle should exhibit this sort of propagation when travels in space-time. This term is vanishingly small compared with the mass term in Equation (31) but very fundamental. Moreover, Our Equations (25) and (27) are equivalent to Dirac equations, Equations (34) and (35), if we replace m_0 by $-im_0$.

Consider now the case when ψ is space independent so that Equation (31) becomes,

$$\frac{d^2\psi}{dt^2}+2\left(\frac{m_0c^2i\beta}{\hbar}\right)\frac{d\psi}{dt}-\left(\frac{m_0c^2}{\hbar}\right)^2\psi=0. \tag{37}$$

with $\psi=\begin{pmatrix}\psi_+\\ \psi_-\end{pmatrix}$ this yields the two equations

$$\frac{d^2\psi_+}{dt^2}+2\left(\frac{m_0c^2i}{\hbar}\right)\frac{d\psi_+}{dt}-\left(\frac{m_0c^2}{\hbar}\right)^2\psi_+=0, \tag{38}$$

And

$$\frac{d^2\psi_-}{dt^2}-2\left(\frac{m_0c^2i}{\hbar}\right)\frac{d\psi_-}{dt}-\left(\frac{m_0c^2}{\hbar}\right)^2\psi_-=0. \tag{39}$$

These two equations have an oscillatory behavior, i.e.,

$$\psi_+(t)=A_+\exp(-i\omega t),\quad \psi_-(t)=A_-\exp(i\omega t), \tag{40}$$

where $\omega=\frac{m_0c^2}{\hbar}$ and A_+, A_- are constants. This means that the particle with the wavefunction ψ has two energy eigen states, one for a particle and the other one for an antiparticle. Hence, Equation (40) reveals that the particle is described by a standing wave having positive and negative energy. This is the essence of Dirac's theory. The two states are separated by an amount of energy, $\Delta E' = 2m_0c^2$.

Using Equation (29), Equation (31) can be written in the form,

$$\frac{1}{c^2}\frac{\partial^2\psi}{\partial t^2} - \nabla^2\psi - 2\left(\frac{m_0ci}{\hbar}\right)\beta\boldsymbol{\alpha}\cdot\nabla\psi + \left(\frac{m_0c}{\hbar}\right)^2\psi = 0, \tag{41}$$

this can be written as

$$\frac{1}{c^2}\frac{\partial^2\psi}{\partial t^2} - \nabla'^2\psi = 0, \tag{42}$$

Where

$$\nabla' = \nabla + i\frac{m_0c\beta}{\hbar}\boldsymbol{\alpha}. \tag{43}$$

Equation (42) is a wave equation in the new coordinate defined by Equation (43). Equation (43) can be written as

$$\boldsymbol{p}' = \boldsymbol{p} + m_0c\beta\boldsymbol{\alpha}. \tag{44}$$

This can be compared with the covariant derivative that results from the interaction of a particle with a photon field A, viz., $\boldsymbol{p}' = \boldsymbol{p} - e\boldsymbol{A}$. Equation (32) can be written as

$$E' = E - m_0c^2\beta. \tag{45}$$

Equations (32) and (43) can be combined into a single equation as

$$D_\mu = \partial_\mu + i\frac{m_0c}{\hbar}\gamma_\mu, \quad \gamma_\mu = (\beta, \beta\alpha). \tag{46}$$

We call here the derivative D'_μ the spinor derivative. With this derivative the Dirac equation takes the simple forms,

$$P\psi = 0, \quad \text{where} \quad P_\mu = i\hbar\gamma^\mu D_\mu, \tag{47}$$

It is interesting to notice that Equations (47) looks like massless Dirac equation. The second term in Equation (44) represents a self interaction of the particle due to its spin. Since the vector potential is a gauge field, the spin of the particle will accordingly become a gauge quantity. In present case, the electron interacts with its spin that is related to α. This effect represents a self interaction of the particle. The algebra of the P_μ's commutator bracket is

$$\left[P^\mu, P^\nu\right] = 2m_0 c i\sigma^{\mu\nu}, \quad \sigma^{\mu\nu} = \frac{i}{2}\left[\gamma^\mu, \gamma^\nu\right]. \tag{48}$$

Thus, unlike partial derivative, spinor derivatives do not commute. The momenta commute for a massless particle. Equation (41) describes a particle with definite mass which after a characteristic distance of $\ell = \frac{\hbar}{m_0 c}$ becomes a wave as described by Equation (42). Hence, the corpuscular nature of the particle is exhibited after a distance of $X_0 = \frac{2\hbar}{m_0 c}$, and the wave nature after a time of $T_0 = \frac{2\hbar}{m_0 c^2}$. The particle's velocity must be in such a way to reach the next point in the same time required to be in the other state. This requires its velocity to be $\mathbf{v} = \boldsymbol{\alpha} c$. Thus, the particle remains in a continuous dual state (particle + wave). This duality is manifested during a time of T_0 at a distance of X_0. This may usher into a quantization of space and time in units of T_0 and X_0 as fundamental units. With the definition $\mathbf{v} = \boldsymbol{\alpha} c$, Dirac's equation can be written as

$$\frac{\partial}{\partial t} + \mathbf{v} \cdot \nabla = -\frac{i m_0 c^2}{\hbar}\beta = \frac{\mathrm{d}}{\mathrm{d}t} \tag{49}$$

which implies that

$$i\hbar \frac{\mathrm{d}\psi}{\mathrm{d}t} = m_0 c^2 \beta\psi. \tag{50}$$

Hence, Equation (50) is a variant form of Dirac's equation. But since ψ can be written as two-components column, viz., $\psi = \begin{pmatrix} \psi_+ \\ \psi_- \end{pmatrix}$, the above equation implies that

$$i\hbar \frac{\mathrm{d}\psi_+}{\mathrm{d}t} = m_0 c^2 \psi_+, \quad i\hbar \frac{\mathrm{d}\psi_-}{\mathrm{d}t} = -m_0 c^2 \psi_-. \tag{51}$$

This shows that the operator $\hat{A} = i\hbar \frac{\mathrm{d}}{\mathrm{d}t}$ is the rest mass energy operator of the individual spinor components.

The Continuity Equation

Taking the complex conjugate of Equation (31) and multiplying it by ψ once from right, and subtract it from Equation (31) after multiplying it by ψ^* from left, we obtain the continuity equation,

$$\frac{\partial \rho_T}{\partial t} + \nabla \cdot \boldsymbol{J} = 0, \tag{52}$$

Where

$$\rho_T = \rho_S + \rho_{KG}, \rho_S = \beta \psi^* \psi,$$
$$\rho_{KG} = \frac{i\hbar}{2m_0 c^2}\left(\psi^* \frac{\partial \psi}{\partial t} - \psi \frac{\partial \psi^*}{\partial t}\right),$$
$$\boldsymbol{J} = \frac{\hbar}{2m_0 i}\left(\psi^* \nabla \psi - \psi \nabla \psi^*\right). \tag{53}$$

It is understood here that ψ is a spinor, ρ_T is the charge density and $\boldsymbol{J}$ is the current density. It is interesting that Equation (31), obtained from Dirac's equation using the differential operator bracket in Equation (30), yields a continuity equation sharing both the Dirac and Klein-Gordon features of the charge (probability) density. This interplay exists despite the fact that Dirac's equation represents a fermionic particle while Klein-Gordon equation represents a bosonic particle.

6. THE SPACE AND TIME INVARIANCE OF DIRAC'S EQUATION

If we apply the transformations in Equations (32) and (43) to Equation (29), Dirac's equation will be invariant. Thus, the space and time transformation represented by Equations (32) and (43) ushers into a new transformation of Dirac's equation that were never known before. With some scrutiny, we know from the theory of relativity that the kinetic energy (E_K) is related to the total energy (E) by $E_K = E - m_0c^2$. In quantum mechanics,

$E \rightarrow i\hbar\frac{\partial}{\partial t}$, so that

$$\hat{E}_K = i\hbar\frac{\partial}{\partial t} - m_0c^2 = i\hbar\left(\frac{\partial}{\partial t} + i\frac{m_0c^2}{\hbar}\right) = i\hbar\frac{\partial}{\partial \eta}. \tag{54}$$

This is the relativistic kinetic energy operator. Alternatively, using Equation (29), this can be written as

$$\hat{E}_K = c\boldsymbol{\alpha}\cdot\hat{\boldsymbol{p}}. \tag{55}$$

This equation implies that Dirac's equation can be obtained from the relativistic energy equation

$$E = E_K + \beta m_0c^2. \tag{56}$$

This equation suggests that there are two possible energy equations. These are $E_{\pm} = E_K \pm m_0c^2$. Hence, a Dirac's particle has in principle two energies,

$E_+ = E_K + m_0c^2$ and $E_- = E_K - m_0c^2$.

Using Equation (32), Dirac's equation Equation (29), becomes

$$\frac{\partial\psi}{\partial\eta}+\nabla\cdot(c\boldsymbol{\alpha})\psi=0. \tag{57}$$

Thus, in the time coordinate η, Dirac's equation represents a continuity-like equation. However, in the real time, the continuity equation in Dirac's formalism, is defined as

$$\frac{\partial\rho}{\partial t}+\nabla\cdot\boldsymbol{J}=0, \tag{58}$$

where $\rho=\psi^{+}\psi$ and $\boldsymbol{J}=\psi^{+}c\boldsymbol{\alpha}\psi$. Using Equation (43), Dirac's equation is transformed into a continuity-like equation in the new space coordinate, viz.,

$$\frac{\partial\psi}{\partial t}+\nabla'\cdot(c\boldsymbol{\alpha})\psi=0. \tag{59}$$

Notice here that $\boldsymbol{J}$ has the same form in both coordinates.

7. SPACE AND TIME INVARIANCE OF MAXWELL'S EQUATIONS

We would like here to apply the space and time transformations in Equations (32) and (43) to explore their implications in Maxwell's equations. These transformations leave Dirac's equation invariant. We know that quantum electrodynamics incorporates the interaction of an electron with a photon. Quantum electrodynamics becomes invariant under gauge transformation, if we replace the partial derivative with a covariant derivative incorporateing the photon field. Analogously, we assume here that Maxwell's equations are invariant under the new space and time transformations in Equations (32) and (43). Applying Equations (32) and (43) to the Ampere's and Faraday's equations [12].

$$\nabla\times\boldsymbol{E}=-\frac{\partial\boldsymbol{B}}{\partial t}, \tag{60}$$

and

$$\nabla \times \boldsymbol{B} = \mu_0 \boldsymbol{J} + \frac{1}{c^2}\frac{\partial \boldsymbol{E}}{\partial t}, \tag{61}$$

yield

$$\boldsymbol{B} = \frac{\boldsymbol{v} \times \boldsymbol{E}}{c^2}, \quad \boldsymbol{v} = c\boldsymbol{\alpha}, \tag{62}$$

and

$$\boldsymbol{E} = -\boldsymbol{v} \times \boldsymbol{B}, \quad \boldsymbol{v} = c\boldsymbol{\alpha}. \tag{63}$$

The remaining two Maxwell's equations

$$\nabla \cdot \boldsymbol{E} = \frac{\rho}{\varepsilon_0}, \quad \nabla \cdot \boldsymbol{B} = 0. \tag{64}$$

yield

$$\boldsymbol{v} \cdot \boldsymbol{E} = 0, \quad \boldsymbol{v} \cdot \boldsymbol{B} = 0, \quad \boldsymbol{v} = c\boldsymbol{\alpha}. \tag{65}$$

It is interesting to notice that Equation (65) is compatible with Equations (62) and (63). Moreover, Equations (62) and (63) define the relations between the electric and magnetic fields produced by the moving charge. If the electric (magnetic) field is known, one can obtain the corresponding magnetic (electric) field. Equation (63) shows that the charge moving with constant velocity experiences no net force. The electric field lines of a moving charge crowded in the direction perpendicular to $\boldsymbol{v}$ and are given by [13]

$$\boldsymbol{E} = \frac{1}{4\pi\varepsilon_0}\frac{q\left(1-\beta^2\right)}{\left(1-\beta^2\sin^2\theta\right)^{\frac{3}{2}}}\frac{\boldsymbol{r}}{r^3}, \quad \beta = \frac{v}{c}, \tag{66}$$

And

$$B = \frac{\mu_0}{4\pi} \frac{q(1-\beta^2)}{(1-\beta^2 \sin^2\theta)^{\frac{3}{2}}} \frac{v \times r}{r^3}, \quad \beta = \frac{v}{c}. \tag{67}$$

Equation (66) shows that the electric and magnetic fields of a moving charge in the forward direction ($\theta = 0$) are less than the electric field of stationary charge. However, the electric and magnetic fields in the perpendicular direction ($\theta = \pi/2$) are bigger than the electric field of stationary charge. Equations (66) and (67) give the relations between the electric and magnetic fields produced by a moving charged particle with constant velocity v. This velocity is given by $v = c\alpha$. This coincides with the quantum mechanics definition of the particle velocity [1-3]. Equation (65) shows that the electric and magnetic fields produced by the charged particle are always perpendicular to the particle's direction of motion. Equation (67) gives, at low velocity, the Biot-Savart law. The power delivered by the fermionic charged particle by its electric and magnetic fields is given by

$$P = F \cdot v = qv \cdot E + qv \cdot (v \times B) = 0$$

The application of the transformations (32) and (43) in the generalized continuity Equations (6) and (8) yields

$$v \cdot J = -\rho c^2, \quad J = -\rho v, \quad v = c\alpha. \tag{68}$$

Since in Dirac formalism $\rho > 0$, the current ushers in a direction opposite to the velocity direction. Moreover, for a constant velocity, one has $(\nabla\rho) \times v = 0$. Equation (68) is very interesting since it defines the charge density (scalar) in terms of the current density (vector). Accordingly, one can define the four vectors in terms of vectorial quantities only.

8. CONCLUDING REMARKS

By introducing three vanishing differential commutator brackets for spinor fields, we have derived a variant form of Dirac's and Klein-Gordon wave equations. Dirac's equation yields a modified Klein-Gordon wave equation. This equation yields directly two energy states for the particle in question. Moreover, Dirac's equation is found to be similar to the continuity equation. We have found time and space transformations under which Dirac's and Maxwell's equations are invariant. In terms of these coordinates, Dirac and Klein-Gordon equations describe a mass-less particle. The invariance of these transformations under Maxwell's equations shows that the electric and magnetic fields produced by a moving charge are perpendicular to velocity of the particle. Hence, there is no power associated with these fields. The space and time transformations show that space and time are quantized in terms of characteristic units of T_0 and X_0. The fermionic charged particle exhibits its wave and corpuscular nature on periodic space and time basis.

ACKNOWLEDGEMENTS

This work is supported by the university of Khartoum research fund. We gratefully acknowledge this support. The critical and useful comments by the anonymous referees are highly acknowledged.

REFERENCES

1. J. D. Bjorken and S. D. Drell, "Relativistic Quantum Mechanics," McGraw-Hill, New York, 1964.
2. L. D. Landau and E. M. Liftshiz, "Quantum Mechanics," 3rd Edition, Pergamon Press, Oxford, 1977.

3. V. B. Berestetskii, L. P. Pitaevskii and E. M. Lifshitz, "Quantum Electrodynamics," 2nd Edition, Vol. 4, Elsevier, Amsterdam, 1982.

4. I. Arbab and Z. Satti, "On the Generalized Maxwell Equations and Their Prediction of Electroscalar Wave," Progress in Physics, Vol. 2, No. 8, 2009, pp. 8-13.

5. I. Arbab and H. M. Widatallah, "The Generalized Continuity Equations," Chinese Physics Letters, Vol. 27, No. 8, 2010, Article ID 084703.

6. I. Arbab and F. A. Yassein, "A New Formulation of Electromagnetism," Journal of Electromagnetic Analysis and Applications, Vol. 2, No. 8, 2010, p. 457.

7. I. Arbab, "A Quaternionic Quantum Mechanics," Applied Physics Research, Vol. 3, No. 2, 2011, p. 160.

8. P. G. Tait, "An Elementary Treatise on Quaternions," 2nd Edition, Cambridge University Press, Cambridge, 1873.

9. H. F. Harmuth, T. W. Barrett and B. Meffert, "Modified Maxwell Equations in Quantum Electrodynamics," World Scientific, River Edge, 2001.

10. G. Feinberg, "Possibility of Faster-Than-Light Particles," Physical Reviews, Vol. 159, No. 5, 1967, pp. 1089-1105.

11. J. Ciborowski, "Hypothesis of Tachyonic Neutrinos," Acta Physicsa Polonica B, Vol. 29, No. 1-2, 1998, pp. 113-121.

12. J. D. Jackson, "Classical Electrodynamics," 2nd Edition, Wiley, New York, 1975.

13. F. Zbigniew, "Lecture Notes in Electromagnetic Theory," University of Queensland, Brisbane, 2005.

2

FRACTAL APPROXIMATION OF MOTION AND ITS IMPLICATIONS IN QUANTUM MECHANICS

Maricel Agop[1,2], Daniela Magop[3], Elena Simona Bacaita [1,4]

[1]Physics Department, Faculty of Machine Manufacturing and Industrial Management, "Gheorghe Asachi" Technical University, Iasi, Romania

[2]Lasers, Atoms and Molecules Physics Laboratory, University of Science and Technology, Lille, France

[3]Physics Department, "Al. I. Cuza" University, Iasi, Romania

[4]Department of Natural and Synthetic Polymers, Faculty of Chemical Engineering and Environmental Protection, "Gheorghe Asachi" Technical University, Iasi, Romania

ABSTRACT

Inconsistencies of some standard quantum mechanical models (Madelung's, de Broglie's models) are eliminated assuming the micro particle movements on continuous, but non-differentiable curves (fractal curves). This hypothesis, named by us the fractal approximation of motion, will allow an unitary approach of the phenomena in quantum mechanics (separation of the physical motion of objects in wave and particle components depending on the scale of resolution, correlated motions of the wave and particle, i.e. wave-particle duality, the mechanisms of duality, by means of both phase wave-particle coherence and wave-particle incoherence, the particle as a clock, particle incorporation into the wave and the implications of such a process). Moreover, correspondences with standard gravitational models (Einstein's model, string theory) can be also distinguished.

KEYWORDS

Fractals; Quantum Mechanics; Gravitational Models

1. INTRODUCTION

We perform a critical analysis of some quantum mechanical models such as the hydrodynamic model (Madelung's model), de Broglie's theory of double solution etc., specifying both mathematical and physical inconsistencies that occur in their construction.

These inconsistencies are eliminated by means of the fractal approximation of motion (physical objects moving on continuous and non-differentiable curves, i.e. fractal curves) developed in the framework of Scale Relativity (SR) [1-7]. The following original results are obtained: 1) Separation of the physical motion of objects in wave and particle components depending on the scale of resolution (differentiable as waves and non-differentiable as particles)—see Sections 5-7; 2) Correlated motions of the wave and particle (wave-particle duality)—see Section 8; 4) The mechanisms of duality (in phase wave-particle coherence, Sections 9 and 10; and 5) Wave-particle incoherence, see Section 11; 6) The particle as a clock, its incorporation into the wave and the implications of such a process—see Sections 12 and 13.

The original results of this work are published in references [8-12]. Explicitly, Eulerian's approximation of motions on fractal curves is presented in [11,12], the hydrodynamic model in a second order approximation of motion in [8], wave-particle duality for "coherent" fractal fluids with the explanation of the potential gap in [9], the physical self-consistence of wave-particle duality in various approximations of motion and for various fractal curves in [11,12]. A unitary treatment of both the problems listed above and their various mathematical and physical extensions are developed in [10].

2. HYDRODYNAMIC MODEL OF QUANTUM MECHANICS (MADELUNG'S MODEL)

Quantum mechanics is substantiated by the Schrődinger wave equation [13-16]

$$i\hbar\frac{\partial\Psi}{\partial t}=\mathbf{U}\Psi-\frac{\hbar^2}{2m_0}\Delta\Psi \tag{1}$$

where $\hbar$ is the reduced Planck's constant, m_0 the rest mass of the test particle, U the external scalar field and Ψ the wave-function associated to the physical system. This differential equation is linear and complex.

Starting from this equation, Madelung [17,18] constructed the following model. One separates real and imaginary parts by choosing Ψ of the form:

$$\Psi(\mathbf{r},t)=R(\mathbf{r},t)e^{iS(\mathbf{r},t)} \tag{2}$$

which induces the velocity field:

$$v=\frac{\hbar}{m_0}\nabla S \tag{3}$$

and the density of the probability field:

$$\rho(\mathbf{r},t)=R^2(\mathbf{r},t) \tag{4}$$

Using these fields one gets the hydrodynamic version of quantum mechanics (Madelung's model)

$$\frac{\partial}{\partial t}(m_0 \rho v) + \nabla(m_0 \rho v v) = -\rho \nabla(\mathbf{U} + \mathbf{Q}) \tag{5}$$

$$\frac{\partial \rho}{\partial t} + \nabla \cdot (\rho v) = 0 \tag{6}$$

Where

$$\mathbf{Q} = -\frac{\hbar^2}{2m_0} \frac{\Delta\sqrt{\rho}}{\sqrt{\rho}} \tag{7}$$

is called the quantum potential. Equation (5) corresponds to the momentum conservation law and Equation (6) to the conservation law of the probability's density field (quantum hydrodynamics equations).

We have the following:

1) Any micro-particle is in constant interaction with an environment called "sub-quantum medium" through the quantum potential Q;

2) The "sub-quantum medium" is identified with a non-relativistic quantum fluid described by the equations of quantum hydrodynamics. In other words, the propagation of the Ψ field from wave mechanics is replaced by a fictitious fluid flow having the density ρ and the speed v, the fluid being in a field of forces $\nabla(\mathbf{U}+\mathbf{Q})$. Moreover, the following model of particle states [19-26], Madelung type fluid in "interaction" with its own "shell" (there is no space limitation of the fluid, though of the particle).

3. DE BROGLIE'S THEORY OF DOUBLE SOLUTION. THE NEED FOR INTRODUCING THE MODEL OF BOHM AND VIGIER

One of the key observations that de Broglie left in the development of quantum mechanics, is the difference between the relativistic transformation of the frequency of a wave and that of a clock's frequency [27-32]. It is well known that, if υ_0 is the frequency of a clock in its own framework, the frequency conferred by an observer who sees it passing with the speed $v = \beta c$ is $\nu_c = \nu_0\sqrt{1-\beta^2}$

This is what is called the phenomenon of "slowing down of horologes". This phenomenon takes place due to the relative motion of horologes. On the contrary, if a wave within a certain reference system is a stationary one, with frequency υ_0 and is noticed in a reference system animated with speed $v = \beta c$, as compared with the first one, it will appear as a progressive wave that propagates in the sense of the relative motion, with frequency

$$\nu = \frac{\nu_0}{\sqrt{1-\beta^2}}$$

and with the phase speed

$$v = \frac{c}{\beta} = \frac{c^2}{v}$$

If the corpuscle,. according to relation W = hv, is given an internal frequency

$$\nu_0 = \frac{m_0 c^2}{h}$$

and if we admit that within the appropriate system of the corpuscle the associated wave is a stationary one, with frequency υ_0, all the fundamental relations of wave mechanics and in particular $\lambda = h/p$, in which p is the impulse of the corpuscle, are immediately obtained from the previous relations.

Since de Broglie considers that the corpuscle is constantly located in the wave, he notices the following consequence: the motion of the corpuscle has such a nature that it ensures the permanent concordance between the phase of the surrounding wave and the internal phase of the corpuscle considered as a small horologe. This relation can be immediately verified in the simple case of a corpuscle in uniform motion, accompanied by a monochromatic plain wave. Thus, when the wave has the general form

$$\Psi = \mathbf{A}(x,y,z,t)e^{\frac{2\pi i}{h}\Phi(x,y,z,t)}$$

in which A and Φ are real, the phase concordance between the corpuscle and its wave requires that the speed of the corpuscle in each point of its trajectory be given by the relation

$$v = -\frac{1}{m_0}\nabla\Phi$$

Nevertheless it was not enough to superpose the corpuscle with the wave, imposing it to be guided by the propagation of the wave: the corpuscle had to be represented as being incorporated in the wave, i.e. as being a part of the structure of the wave. De Broglie was thus directed to what he himself called the theory of "double solution". This theory admits that the real wave is not a homogeneous one, that it has a very small area of high concentration of the field that represents the corpuscle and that, besides this very small area, the wave appreciably coincides with the homogeneous wave as formulated by the usual wave mechanics.

The phenomenon of guiding the particle by the surrounding wave field results from the fact that the equations of the field are not linear ones and that this lack of linearity, that almost exclusively shows itself in the corpuscular area, correlating the motion of the particle with the propagation of the surrounding wave [30-33].

Nevertheless there is a consequence of "guidance" upon which we should insist. Even if a particle is not submitted to any external field, if the wave that surrounds it is not an appreciably plain and monochromatic one (therefore if this wave has to be represented through a superposition of monochromatic plain waves) the motion that the guidance formula imposes is not rectilinear and uniform. The corpuscle is subjected by the surrounding wave, to a force that curves its trajectory: this "quantum force" equals the gradient with the changed sign of the quantum potential Q given by (7). Therefore, the uniform motion of the wave has to be superposed with a "Brownian" motion having random character that is specific to the corpuscle.

Under the influence of Q, the corpuscle, instead of uniformly following one of the trajectories that are defined by the guidance law, constantly jumps from one of these trajectories to another, thus passing in a very short period of time, a considerably big number of sections within these trajectories and, while the wave remains isolated in a finite area of the space, this zigzag trajectory hurries to explore completely all this region. In this manner, one can justify that the probability of the particle to be present in a volume element $d\tau$ of the physical space is equal to $|\Psi|^2 d\tau$. This is what Bohm and Vigier did in their statement: therefore they showed that the probability of repartition in $|\Psi|^2$ must take place very quickly. The success of this demonstration must be correlated with the characteristics if "Markov's chains" [19,34-36].

4. COMMENTS

In his attempt to built the theory of the double solution, de Broglie admits certain assertions [27-32]: 1) The frequency of the corpuscle that is assimilated to a small horologe must be identified with the frequency of the associated progressive wave; 2) The coherence of the inner phase of the corpuscle-horologe with the phase of the associated wave; 3) The corpuscle must be "incorporated" into the progressive associated wave through the "singularity" state. Thus, the motion of the corpuscle "correlates" with the propagation of the associated progressive wave. Nevertheless, once we admit these statements, de Broglie's theory does not answer a series of problems, such as, for example: 1) What are the mechanisms through which either the wave feature or the corpuscular one impose, either both of them in the stationary case as well as in the non stationary one? 2) The limits in the wave-corpuscle system of the corpuscular component as well as the wave one and their correspondence; 3) How is the “solidarity” between the motion of the corpuscle and the one of the associated progressive wave naturally induced? 4) What are the consequences of this “solidarity”? Others discrepancies are mentioned in [27- 32].

Moreover, Madelung’s theory [17,18] brings new problems. How can we built a pattern of a corpuscle (framework + Madelung liquid) endlessly extended in space?

Here are some of the “drawbacks” of the patterns in Sections 2 and 3 which we shall analyze and remove by means of introducing the fractal approximation of the motion.

5. THE MOTION EQUATION OF THE PHYSICAL OBJECT IN THE FRACTAL APPROXIMATION OF MOTION. THE EULERIAN SEPARATION OF MOTION ON RESOLUTION SCALES

The fractal approximation of motion refers to the movement of physical objects (wave + corpuscle) on continuous and non differentiable curves (fractal curves). This approximation is based on the scale Relativity theory (RS) [1-7]. Thus, the fractal differential operator can be introduced

$$\frac{\hat{\mathrm{d}}}{\mathrm{d}t} = \frac{\partial}{\partial t} + \hat{\mathbf{V}} \cdot \nabla - i\frac{\lambda^2}{2\tau}\left(\frac{\mathrm{d}t}{\tau}\right)^{(2/D_F)-1} \tag{8}$$

where $\hat{\mathbf{V}}$ is the complex speed field

$$\hat{\mathbf{V}} = \mathbf{V} - i\mathbf{U} \tag{9}$$

λ is the scale length, dt is the temporary resolution scale, τ is the specific time to fractal-non fractal transition, and D_F is the arbitrary and constant fractal dimension. Regarding the fractal dimension, we can use any of Hausdorff-Bezicovici, Minkowski-Bouligand or Kolmogoroff dimensions, etc. [3,5-7,36-46]. The only restriction refers to the maintaining of the same type of fractal dimension during the dynamic analysis. The real part of the speed field $\mathbf{V}$ is differentiable and independent as compared with the resolution scale, while the imaginary scale $\mathbf{U}$ is non differentiable (fractal) and depends on the resolution scale.

Now we can apply the principle of scale covariance by substituting the standard time derivate $(\mathrm{d}/\mathrm{d}t)$ with the complex operator $\hat{\mathrm{d}}/\mathrm{d}t$. Accordingly, the equation of fractal space-time geodesics (the motion equation in second order approximation, where second order derivates are used) in a covariant form:

$$\frac{\mathrm{d}\hat{\mathbf{V}}}{\mathrm{d}t} = \frac{\partial \hat{\mathbf{V}}}{\partial t} + \hat{\mathbf{V}} \cdot \nabla \hat{\mathbf{V}} - i \frac{\lambda^2}{2\tau}\left(\frac{\mathrm{d}t}{\tau}\right)^{(2/D_F)-1} \nabla^2 \hat{\mathbf{V}} = 0 \tag{10}$$

This means that the sum of the local acceleration $\partial \hat{\mathbf{V}}/\partial t$, convection $\hat{\mathbf{V}} \cdot \nabla \hat{\mathbf{V}}$ and "dissipation" $\nabla^2 \hat{\mathbf{V}}$ reciprocally compensate in any point of the arbitrarily fractal chosen trajectory of a physical object.

Formally, (10) is a Navier-Stokes type equation, with an imaginary viscosity coefficient,

$$\eta = i \frac{\lambda^2}{2\tau}\left(\frac{\mathrm{d}t}{\tau}\right)^{(2/D_F)-1} \tag{11}$$

This coefficient depends on two temporary scales, as well as on a length scale. The existence of a pure imaginary structured coefficient specifies the fact that "the environment" has rheological features (viscoelastic and hysteretic ones [47-49].

For

$$\frac{\lambda^2}{2\tau}\times\left(\frac{dt}{\tau}\right)^{(2/D_F)-1}\rightarrow 0 \tag{12}$$

Equation (10) reduces to Euclidian form [10-12]:

$$\frac{\partial \hat{\mathbf{V}}}{\partial t}+\hat{\mathbf{V}}\cdot\nabla\hat{\mathbf{V}}\equiv 0 \tag{13}$$

and, hence, separating the real part from the imaginary one

$$\frac{\partial \mathbf{V}}{\partial t}+\mathbf{V}\cdot\nabla\mathbf{V}-\mathbf{U}\cdot\nabla\mathbf{U}=0 \tag{14a}$$

$$\frac{\partial \mathbf{U}}{\partial t}+\mathbf{U}\cdot\nabla\mathbf{V}+\mathbf{V}\cdot\nabla\mathbf{U}=0 \tag{14b}$$

Equation (14a) corresponds to the law of the impulse conservation at differentiable scale (the undulatory component), while (14b) corresponds to the same law, but at a non differentiable scale (corpuscular component). As we will later show, in the case of irrotational movements (14) it will be assimilated to the law of mass conservation.

6. ROTATIONAL MOTIONS AND FLOW REGIMES OF A FRACTAL FLUID

For rotational motions, $\nabla\times\hat{\mathbf{V}}\neq 0$relation (10) with (9) through separating the real part from the imaginary one, i.e. through separating the motions at a differential scale (undulatory characteristic) and non differential one (corpuscular characteristic), results [10].

$$\frac{\partial \mathbf{V}}{\partial t}+\mathbf{V}\cdot\nabla\mathbf{V}-\mathbf{U}\cdot\nabla\mathbf{U}-\frac{\lambda^2}{2\tau}\left(\frac{dt}{\tau}\right)^{(2/D_F)-1}\Delta\mathbf{U}=0 \tag{15a}$$

$$\frac{\partial \mathbf{U}}{\partial t}+\mathbf{U}\cdot\nabla\mathbf{V}+\mathbf{V}\cdot\nabla\mathbf{U}+\frac{\lambda^2}{2\tau}\left(\frac{dt}{\tau}\right)^{(2/D_F)-1}\Delta\mathbf{V}=0 \tag{15b}$$

According to the operator relations

$$\mathbf{V}\cdot\nabla\mathbf{V}=\nabla\left(\frac{\mathbf{V}^2}{2}\right)-\mathbf{V}\times(\nabla\times\mathbf{V}) \tag{16a}$$

$$\mathbf{U}\cdot\nabla\mathbf{U}=\nabla\left(\frac{\mathbf{U}^2}{2}\right)-\mathbf{U}\times(\nabla\times\mathbf{U}) \tag{16b}$$

$$\mathbf{U}\cdot\nabla\mathbf{V}+\mathbf{V}\cdot\nabla\mathbf{U}=\nabla(\mathbf{U}\cdot\mathbf{V})-\mathbf{V}\times(\nabla\times\mathbf{U})-\mathbf{U}\times(\nabla\times\mathbf{V}) \tag{16c}$$

Equations (15) take equivalent forms

$$\frac{\partial \mathbf{V}}{\partial t}+\nabla\left(\frac{\mathbf{V}^2}{2}-\frac{\mathbf{U}^2}{2}\right)-\mathbf{V}\times(\nabla\times\mathbf{V})$$
$$-\mathbf{U}\times(\nabla\times\mathbf{U})-\frac{\lambda^2}{2\tau}\left(\frac{dt}{\tau}\right)^{(2/D_F)-1}\Delta\mathbf{U}=0 \tag{17a}$$

$$\frac{\partial \mathbf{U}}{\partial t}+\nabla(\mathbf{V}\cdot\mathbf{U})-\mathbf{V}\times(\nabla\times\mathbf{U})$$
$$-\mathbf{U}\times(\nabla\times\mathbf{V})+\frac{\lambda^2}{2\tau}\left(\frac{dt}{\tau}\right)^{(2/D_F)-1}\Delta\mathbf{V}=0 \tag{17b}$$

We can now characterize the flow regimes of the fractal fluid at different scales, using some classes of Reynolds numbers. At a differential scale we have

$$R(\text{differential}-\text{nondifferential})=R(D-N)$$
$$=\frac{|\mathbf{V}\cdot\nabla\mathbf{V}|}{D|\nabla\mathbf{U}|}\approx\frac{\mathbf{V}^2 l^2}{DUL} \tag{18}$$

$$R(\text{differential} - \text{nondifferential}) = R(D - N) = \frac{|\mathbf{U} \cdot \nabla \mathbf{U}|}{D|\nabla \mathbf{U}|} \approx \frac{\mathbf{U}l}{D} \tag{19}$$

with

$$D = \frac{\lambda^2}{2\tau}\left(\frac{\mathrm{d}t}{\tau}\right)^{(2/D_F)-1} \tag{20}$$

and at nondifferential scale.

$$R(\text{differential} - \text{nondifferential} - \text{differential transition}) = R(\text{TDN} - D) = \frac{|\mathbf{U} \cdot \nabla \mathbf{V}|}{D|\Delta \mathbf{V}|} \approx \frac{\mathbf{U}l}{D} \tag{21}$$

$$R(\text{differential} - \text{nondifferential} - \text{differential transition}) = R(\text{TDN} - D) = \frac{|\mathbf{V} \cdot \nabla \mathbf{U}|}{D|\Delta \mathbf{V}|} \approx \frac{\mathbf{U}L^2}{Dl} \tag{22}$$

In previous relations V, L, D, are the specific parameters, while U, l, D are the parameters of the nondifferential scale. The parameters V, U are specific speeds, L, l specific lengths and D is a viscosity coefficient. Moreover, the common “element” for R(D ‾ N), R(N ‾ N), R(TDN ‾ D) and R(TND ‾ D) is the “viscosity” which, through (20) is imposed by the resolution scale.

Equations (15) are simplified in the case of the stationary motion for small Reynolds numbers. Thus, Equation (15) for small R(D ‾ N) becomes

$$-\mathbf{U} \cdot \nabla \mathbf{U} - \frac{\lambda^2}{2\tau}\left(\frac{\mathrm{d}t}{\tau}\right)^{(2/D_F)-1} \nabla \mathbf{U} = 0 \tag{23}$$

and for small R(N ‾ N)

$$-\mathbf{V}\cdot\nabla\mathbf{V}-\frac{\lambda^2}{2\tau}\left(\frac{dt}{\tau}\right)^{(2/D_F)-1}\nabla\mathbf{U}=0 \tag{24}$$

Equation (15b) for small R(TDN ¯ D) takes the form

$$\mathbf{V}\cdot\nabla\mathbf{U}+\frac{\lambda^2}{2\tau}\left(\frac{dt}{\tau}\right)^{(2/D_F)-1}\Delta\mathbf{V}=0 \tag{25}$$

and for small R(TND ¯ D)

$$\mathbf{U}\cdot\nabla\mathbf{V}+\frac{\lambda^2}{2\tau}\left(\frac{dt}{\tau}\right)^{(2/D_F)-1}\Delta\mathbf{V}=0 \tag{26}$$

7. IROTATIONAL MOTIONS OF A FRACTAL FLUID. THE INCORPORATION OF THE ASSOCIATE WAVE CORPUSCLE THROUGH THE SOLIDARITY OF MOVEMENTS AND GENERATION OF SCHRODINGER EQUATION

For irrotational motions

$$\nabla\times\hat{\mathbf{V}}=0 \tag{27}$$

which implies

$$\nabla\times\mathbf{V}=0,\nabla\times\mathbf{U}=0 \tag{28}$$

$$\frac{\partial\mathbf{V}}{\partial t}+\nabla\left(\frac{\mathbf{V}^2}{2}\right)-i\frac{\lambda^2}{2\tau}\left(\frac{dt}{\tau}\right)^{(2/D_F)-1}\Delta\mathbf{V}=0 \tag{29}$$

Since through (27) the complex speed field is expressed by means of a scalar function gradient F,

$$\hat{\mathbf{V}} = \nabla\Phi \tag{30}$$

Equation (29) taking into account the operator identities

$$\frac{\partial}{\partial t}\nabla = \nabla\frac{\partial}{\partial t}, \quad \nabla\Delta = \Delta\nabla \tag{31}$$

takes the form

$$\nabla\left[\frac{\partial\Phi}{\partial t} + \frac{1}{2}(\nabla\Phi)^2 - \mathrm{i}\frac{\lambda^2}{2\tau}\left(\frac{\mathrm{d}t}{\tau}\right)^{(2/D_F)-1}\Delta\Phi\right] = 0 \tag{32}$$

or furthermore, through integration

$$\frac{\partial\Phi}{\partial t} + \frac{1}{2}(\nabla\Phi)^2 - i\frac{\lambda^2}{2\tau}\left(\frac{\mathrm{d}t}{\tau}\right)^{(2/D_F)-1}\Delta\Phi = F(t) \tag{33}$$

where F(t) is an arbitrary function depending only on time.

In particular, for F having the form

$$\Phi = -2i\frac{\lambda^2}{2\tau}\left(\frac{\mathrm{d}t}{\tau}\right)^{(2/D_F)-1}\ln\Psi \tag{34}$$

where Y is a new complex scalar function, Equation (46), with the operator identity

$$\frac{\Delta\Psi}{\Psi} = \Delta\ln\Psi + (\nabla\ln\Psi)^2 \tag{35}$$

takes the form:

$$\frac{\lambda^4}{4\tau^2}\left(\frac{\mathrm{d}t}{\tau}\right)^{(4/D_F)-2}\Delta\Psi + i\frac{\lambda^2}{2\tau}\left(\frac{\mathrm{d}t}{\tau}\right)^{(2/D_F)-1}\frac{\partial\Psi}{\partial t} + \frac{F(t)}{2}\Psi = 0 \tag{36}$$

The Schrodinger "geodesics" can be obtained as a particular case of Equation (36), based on the following hypothesis (conditions of solidarity of the motion, incorporating the associated wave corpuscle): 1) The motions of the micro-particles take place on fractal curves with the fractal dimension $D_F = 2$, i.e. the Peano curves [3,5]; 2) $d_{\pm}\xi^i$ are the Markov-Wiener type stochastic variables [3,5] that satisfy the rule

$$\left\langle d_{\pm}\xi^i d_{\pm}\xi^l \right\rangle = \pm\delta^{il}\frac{\lambda^2}{\tau}\mathrm{d}t \tag{37}$$

3) Space scale l and temporary one τ are specific for the Compton scale

$$\lambda = \frac{\hbar}{m_0 c}, \quad \tau = \frac{\hbar}{m_0 c^2} \tag{38}$$

with m_0 the rest mass of the microparticle, c the speed of light in vacuum and $\hbar$ the reduced Planck constant. The parameters (38) should not be understood as "structures" of the standard space-time, but as standards of scale space-time; 4) Function F(t) from (36) is null. Under these circumstances, (36) is reduced to the standard form of Schrodinger's Equation [13,15]

$$\frac{\hbar^2}{2m_0}\Delta\Psi + i\hbar\frac{\partial\Psi}{\partial t} = 0 \tag{39}$$

In such a context, the scale potential of the complex speeds plays the role of the wave function.

8. EXTENDED HYDRODYNAMIC MODEL OF SCALE RELATIVITY AND INCORPORATION OF ASSOCIATED WAVE CORPUSCLE THROUGH FRACTAL POTENTIAL. THE CORRESPONDENCE WITH MADELUNG MODEL

Substituting the complex speed (9) with the restriction (27) and separating the real part with the imaginary one, we obtain the set of differential Equations [10]

$$m_0 \frac{\partial \mathbf{V}}{\partial t} + m_0 \nabla\left(\frac{\mathbf{V}^2}{2}\right) = -\nabla(\mathbf{Q}) \tag{40a}$$

$$\frac{\partial \mathbf{U}}{\partial t} + \nabla(\mathbf{V}\cdot\mathbf{U}) + \frac{\lambda^2}{2\tau}\left(\frac{\mathrm{d}t}{\tau}\right)^{(2/D_F)-1} \Delta\mathbf{V} = 0 \tag{40b}$$

where Q is the fractal potential, expressed as follows:

$$\mathbf{Q} = -\frac{m_0 \mathbf{U}^2}{2} - \frac{m_0}{2}\frac{\lambda^2}{2\tau}\left(\frac{\mathrm{d}t}{\tau}\right)^{(2/D_F)-1} \nabla\cdot\mathbf{U} \tag{41}$$

For

$$\Psi = \sqrt{\rho e^{iS}} \tag{42}$$

with $\sqrt{\rho}$ an amplitude and S a phase, then (34) under the form

$$\Phi = -i\frac{\lambda^2}{\tau}\left(\frac{\mathrm{d}t}{\tau}\right)^{(2/D_F)-1} \ln\left(\sqrt{\rho e^{iS}}\right)$$

implies the complex speed fields of components

$$\mathrm{V} = \frac{\lambda^2}{2\tau}\left(\frac{\mathrm{d}t}{\tau}\right)^{(2/D_F)-1} \nabla S \tag{43a}$$

$$\mathbf{U} = \frac{\lambda^2}{2\tau}\left(\frac{\mathrm{d}t}{\tau}\right)^{(2/D_F)-1} \nabla \ln \rho \tag{43b}$$

From the perspective of Equation (43), the Equation (40) keeps its form, and the fractal potential is given by the simple expression

$$\mathbf{Q} = -m_0 \frac{\lambda^2}{\tau}\left(\frac{\mathrm{d}t}{\tau}\right)^{(2/D_F)-1} \frac{\Delta\sqrt{\rho}}{\sqrt{\rho}} \tag{44}$$

Again through Equations (43), Equation (40b) takes the form:

$$\nabla\left(\frac{\partial \ln \rho}{\partial t} + \mathbf{V}\cdot\nabla \ln \rho + \nabla\cdot\mathbf{V}\right) = 0$$

or, still, through integration with $\rho \neq 0$

$$\frac{\partial \rho}{\partial t} + \nabla\cdot(\rho\mathbf{V}) = T(t) \tag{45}$$

with T(t), an exclusively time dependent function Equation (40) corresponds to the impulse conservation law at differential scale (the classical one), while the impulse conservation law at non differential scale is expressed through (45) with $T(t) \equiv 0$, as a probability density conservation law.

Therefore, equations

$$m_0\left(\frac{\partial \mathbf{V}}{\partial t} + \mathbf{V}\cdot\nabla\left(\frac{\mathbf{V}^2}{2}\right)\right) = -\nabla(\mathbf{Q}) \tag{46a}$$

$$\frac{\partial \rho}{\partial t} + \nabla\cdot(\rho\mathbf{V}) = 0 \tag{46b}$$

with Q given by (41) or (44) forms the set of equations of scale relativity extended hydrodynamics in fractal dimension D_F. We mention that in references [3,5-7] the model has been extended only for $D_F = 2$. The fractal potential (41) or (44) is induced by the non differentiability of space-time.

In an external scalar field U, the system of Equation (46) modifies as follows:

$$m_0\left[\frac{\partial \mathbf{V}}{\partial t}+\nabla\left(\frac{\mathbf{V}^2}{2}\right)\right]=-\nabla(\mathbf{Q}+\mathbf{U}) \tag{47a}$$

$$\frac{\partial \rho}{\partial t}+\nabla(\rho\mathbf{V})=0 \tag{47b}$$

Now the quantum mechanics in hydrodynamic formula (Madelung's model [17]) is obtained as a particular case of relations (47), using the following hypothesis: 1) The motion of the micro-particles takes place on Peano curves with $D_F = 2$; 2) $d_+\xi^i$ are the Markov-Wiener variables [3,5-7]; 3) The time space scale is a Compton one. Then, (38) have the expressions

$$\mathbf{V}=\frac{\hbar}{m_0}\nabla S,\quad \mathbf{U}=\frac{\hbar}{2m_0}\nabla \ln \rho \tag{48}$$

and (41),

$$\mathbf{Q}=-\frac{m_0\mathbf{U}^2}{2}-\frac{\hbar}{2}\nabla\cdot\mathbf{U} \tag{49}$$

9. "MECHANISMS" OF DUALITY THROUGH COHERENCE IN CORPUSCLE-WAVE PHASE

In the stationary case, the system of Equations (46(a) and (b)) become [10]

$$\nabla\left(\frac{\mathbf{V}^2}{2}+\mathbf{Q}\right)=0 \tag{50a}$$

$$\nabla(\rho\mathbf{V})=0 \tag{50b}$$

or, still, through integration

$$\frac{\mathbf{V}^2}{2}+\mathbf{Q}=E=\text{const.} \tag{51a}$$

$$\rho\mathbf{V}=\text{const.} \tag{51b}$$

Let us choose the null power density in (51b). Then there is no impulse transport at differential scale between corpuscle and wave. Moreover, for $\rho\neq 0$.

$$\mathbf{V}=0 \tag{52}$$

which implies through relation (43)

$S=\text{const.}$ (53)

In other words, the fluid becomes coherent (the fluid particles have the same phase). Such a state is specific for quantum fluids [50,51], such as superconductors, super-fluids, etc. [14,52]. Under such circumstances, the phase of the corpuscle considered as a small horologe equals the phase of the associated wave (coherence in corpuscle-wave phase).

At non-differential scale, Equation (51), with restriction (52) takes the form

$$\mathbf{Q} = -\frac{2m_0 D^2 \Delta\sqrt{\rho}}{\sqrt{\rho}} = -\frac{m_0 \mathbf{U}^2}{2} - m_0 D\nabla\mathbf{U} = E = \text{const} \tag{54a}$$

$$\mathbf{D} = \frac{\lambda^2}{\tau}\left(\frac{\mathrm{d}t}{\tau}\right)^{(2/D_F)-1} \tag{54b}$$

or, still, by applying the gradient operator

$$\mathbf{A} = \nabla\left(\sqrt{\rho}\right) \tag{55}$$

$$\Delta\mathbf{A} + \frac{E}{2m_0 D^2}\mathbf{A} = 0 \tag{56}$$

We distinguish the following situations:

1) For E > 0 and with substitution

$$\frac{1}{\Lambda^2} = \frac{E}{2m_0 D^2} \tag{57}$$

Equation (56) becomes

$$\Delta\mathbf{A} + \frac{1}{\Lambda^2}\mathbf{A} = 0 \tag{58}$$

Therefore: a) The space oscillations of field A and, therefore the space associated with the motion of coherent fluid particles is endowed with regular non homogeneities (of lattice type). In other words, the field A crystallizes the space. The one dimensional space "crystal" has the constant of the network

$$\Lambda = \frac{\lambda^2}{2\tau}\left(\frac{\mathrm{d}t}{\tau}\right)^{(2/D_F)-1}\left(\frac{2m_0}{E}\right)^{1/2} \tag{59}$$

that depends both on the "viscosity" – iη given by (11) and on the energy of the particle; b) The one dimensional geodesics of the "crystallized" space given by the expression

$$\rho(x) = \mathbf{A}^2 \sin^2(kx+\delta) \tag{60}$$

implies both fractal speed

$$\mathbf{U}_x = D\frac{\mathrm{d}\ln\rho}{\mathrm{d}x} = 2Dkc\,\mathrm{tg}(kx+\delta) \tag{61}$$

and fractal potential

$$\mathbf{Q}_x = -\frac{m_0\mathbf{U}_x^2}{2} - m_0 D\frac{\mathrm{d}\mathbf{U}_x}{\mathrm{d}x} = -2m_0D^2k^2\mathrm{ctg}^2(kx+\delta) + 2m_0D^2k^2\frac{1}{\sin^2(kx+\delta)} = 2m_0D^2k^2 \tag{62}$$

with A and δ, the integration constants

$$k = \frac{1}{\Lambda} \tag{63}$$

c) For the movements of micro particles on Peano curves ($D_F = 2$) at Compton scale

$$D = 2m_0\hbar$$

Therefore, through (62), under the form

$$\mathbf{Q}_x = 2m_0D^2k^2 = \frac{p_x^2}{2m_0}, \tag{64a}$$

$$p_x = 2m_0Dk \tag{64b}$$

De Broglie "quantum" impulse is found

$$p_x = \frac{\hbar}{\Lambda} \tag{65}$$

d) The dominant of the wave characteristic is achieved by the “self diffraction” mechanism of the fractal field, ρ, on the one dimensional space “crystal” of constant Λ induced by the same field. Indeed, relation (61) with notations

$$\Phi = kx + \delta, \tag{66a}$$

$$k = \frac{1}{\Lambda} \tag{66b}$$

In approximation $\Phi \approx 1$, i.e. for $\mathrm{tg}\Phi \approx \sin\Phi$ and using Nottale’s relation [1-7] $2D/\mathbf{U}_x \approx n\lambda$ it takes the common form (Bragg’s relation)

$$\Lambda \sin\Phi \approx n\lambda \tag{67}$$

This result is in concordance with the recently expressed opinion in [53-55]; e) There is impulse transfer on the fractal field between the corpuscle and the wave; f) According to Taylor’s criterion [56] self-organization (crystallization and self diffraction of the space) appears when the energy of the system is minimal. This can be immediately verified using relation (51a).

2) For E = 0, Equations (51a) and (56) have the same form

$$\Delta\sqrt{\rho}=0 \quad \Delta\mathbf{A} = 0 \tag{68}$$

It follows that: a) The geodesics are expressed through harmonic functions and the particle finds itself in a critical state, i.e. the one that corresponds to the wave-corpuscle transition; b) In the one dimensional case, the geodesics have the form

$$\rho(x) = kx + \delta \tag{69}$$

which induces the fractal speed field

$$\mathbf{U}_x = \frac{D}{kx+\delta} \tag{70}$$

namely the null value of the fractal potential

$$\mathbf{Q}_x = -\frac{m_0}{2}\frac{D^2}{(kx+\delta)^2} + \frac{m_0}{2}\frac{D^2}{(kx+\delta)^2} = 0 \tag{71}$$

c) Although the energy is null, there is impulse transfer between corpuscle and wave on the fractal component of the speed field.

3) For E < 0 and with notations

$$\frac{1}{\overline{\Lambda}^2} = \frac{\overline{E}}{2m_0 D^2}, \quad E = -\overline{E} \tag{72}$$

Equation (56) takes the form

$$\Delta\mathbf{A} - \frac{1}{\overline{\Lambda}^2}\mathbf{A} = 0 \tag{73}$$

The following aspects result: a) Field A is expelled from the structure, its penetration depth being

$$\overline{\Lambda} = \frac{\lambda^2}{2\tau}\left(\frac{dt}{\tau}\right)^{(2/D_F)-1}\left(\frac{2m_0}{\overline{E}}\right)^{1/2} \tag{74}$$

b) The one-dimensional geodesics of the space are described through function

$$\rho(x) = \overline{\Lambda}^2 \mathrm{sh}^2\left(\overline{k}x + \overline{\delta}\right) \tag{75}$$

and lead to the fractal speed

$$\mathbf{U}_x = 2D\overline{k}cth\left(\overline{k}x + \overline{\delta}\right) \tag{76}$$

the fractal potential respectively

$$\mathbf{Q}_x = -2m_0 D^2 \bar{k}^2 \operatorname{cth}^2\left(\bar{k}x + \bar{\delta}\right) + 2m_0 D^2 \frac{\bar{k}^2}{\operatorname{sh}^2\left(\bar{k}x + \bar{\delta}\right)} = -2m_0 D^2 \bar{k}^2 \tag{77}$$

where $\bar{\Lambda}, \bar{\delta}$ are two integration constants

$$\bar{k} = \frac{1}{\bar{\Lambda}} \tag{78}$$

c) The dominant of the corpuscular characteristic is accomplished by means of “self-expulsion” mechanism of the fractal field from its own structure that it generates (that is the corpuscle), the penetration depth being $\bar{\Lambda}$. The identification

$$\mathbf{Q}_x = -2m_0 D^2 \bar{k}^2 = \frac{\bar{p}^2}{2m_0} \tag{79}$$

implies the purely imaginary impulse

$$\bar{p} = -2im_0 D\bar{k} \tag{80}$$

that suggests ultra rapid virtual states (ultra rapid motions in the wave field, resulting in the “singularity” of the field, i.e. the corpuscle). As a matter of fact, if we consider de Broglie’s original theory (motions on Peano curves with $D_F = 2$, at Compton’s scale), singularity (the corpuscle) moves “suddenly” and chaotically in the wave field, the wave-corpuscle coupling being accomplished through the fractal potential. The corpuscle “tunnels” the potential barrier imposed by the field of the associate progressive wave, generating particle-antiparticle type pairs (ghost type fields (Bittner E.R., 2000)). Nevertheless this model cannot specify the type of the physical process by means of which we reach such a situation: it is only the second quantification that can do this [50, 51,54-57]; d) There is an impulse transfer between the corpuscle and the wave on the fractal component of the speed field, so that all the attributes of the differential speed could be transferred on the fractal speed.

All the above results indicate that wave-particle duality is an intrinsic property of space and not of the particle.

10. WAVE-CORPUSCLE DUALITY THROUGH FLOWING STATIONARY REGIMES OF A COHERENT FRACTAL FLUID IN PHASE. THE POTENTIAL WELL

According to the previous paragraph, let us study the particle in a potential well with infinite width and walls. Then the speed complex field has the form [8-12]

$$\hat{\mathbf{V}}_x = \mathbf{V}_x - i\mathbf{U}_x = 0 - 2iD\left(\frac{n\pi}{a}\right)\mathrm{ctg}\left(\frac{n\pi}{a}\right)x \tag{81}$$

and generates the fractal potential (the energy of the structure) under the form of the noticeable

$$\mathbf{Q}_n = 2m_0 D^2\left(\frac{n\pi}{a}\right)^2 = E_n \tag{82}$$

The last relation (82) allows the implementation of Reynold's criterion

$$R(n) = \frac{\mathbf{V}_c a}{D} = 2n\pi, \quad \mathbf{V}_c = \left(\frac{2E_n}{m_0}\right)^{\frac{1}{2}} \tag{83}$$

For movements on Peano curves ($D_F = 2$) at Compton scale $(2m_0 D = \hbar)$ (83) with substitutions

$$m_0\mathbf{V}_c = \Delta P_x, \quad a = \Delta x \tag{84}$$

and n = 1 reduces to Heinsenberg's relation of uncertainty under equal form

$$\Delta p_x \Delta x = \frac{h}{2} \tag{85}$$

while for $n < +\infty$ it implies a Ruelle-Takens' type criterion of evolution towards chaos [58,59]. Therefore, the wave-corpuscle duality is accomplished through the flowing regimes of a fractal fluid that is coherent in phase. Thus, the laminar flow (small n) induces a dominant ondulatory characteristic, while the turbulent flow (big n) induces a dominant corpuscular characteristic.

11. WAVE-CORPUSCLE DUALITY THROUGH NON-STATIONARY REGIMES OF AN INCOHERENT FRACTAL FLUID

In the one dimensional case the equations of hydrodynamics (46) take the form

$$m_0\left(\frac{\partial \mathbf{V}}{\partial t}+\mathbf{V}\cdot\frac{\partial \mathbf{V}}{\partial x}\right)=-\frac{\partial}{\partial x}\left[-2m_0 D^2\frac{1}{\rho^{1/2}}\frac{\partial^2}{\partial x^2}\left(\rho^{1/2}\right)\right] \tag{86a}$$

$$\frac{\partial \rho}{\partial t}+\frac{\partial}{\partial x}\left(\rho \mathbf{V}\right)=0 \tag{86b}$$

Imposing the initial conditions

$$\mathbf{V}\left(x,t=0\right)=c=\text{const} \tag{87a}$$

$$\rho\left(x,t=0\right)=\frac{1}{\pi^{1/2}\alpha}e^{-\left(\frac{x}{\alpha}\right)^2}=\rho_0 \tag{87b}$$

and on the frontier

$$\mathbf{V}(x = ct, t) = c; \tag{88a}$$

$$\rho(x = -\infty, t) = \rho(x = +\infty, t) = 0 \tag{88b}$$

The solutions of the system (86), using the method in [60], have the expressions

$$\rho(x,t) = \frac{1}{\pi^{1/2}\left[\alpha^2 + \left(\frac{2D}{\alpha}t\right)^2\right]^{1/2}} \exp\left[\frac{(x-ct)^2}{\alpha^2 + \left(\frac{2D}{\alpha}t\right)^2}\right] \tag{89a}$$

$$\mathbf{V} = \frac{c\alpha^2 + \left(\frac{2D}{\alpha}\right)^2 tx}{\alpha^2 + \left(\frac{2D}{\alpha}t\right)^2} \tag{89b}$$

The complex speed field is obtained

$$\hat{\mathbf{V}} = \mathbf{V} - i\mathbf{U} = \frac{c\alpha^2 + \left(\frac{2D}{\alpha}\right)^2 tx}{\alpha^2 + \left(\frac{2D}{\alpha}t\right)^2} + 2iD\frac{x - ct}{\alpha^2 + \left(\frac{2D}{\alpha}t\right)^2} \tag{90}$$

and the field of fractal forces

$$F = 4m_0 D^2 \frac{(x - ct)}{\left[\alpha^2 + \left(\frac{2D}{\alpha}t\right)^2\right]^2} \tag{91}$$

Therefore: 1) Both differential scale speed V and nondifferential one U are not homogeneous in x and t. Under the action of fractal force F, the corpuscle is assimilated to the wave, is a part of its structure, so that it joins the movement of the corpuscle with the propagation of the associated progressive wave; 2) The timing of the movements at the two scales, V = U implies the space-time homographic dependence

$$x = \frac{c\alpha^2}{2D} \frac{1 + \frac{2D}{\alpha^2} t}{1 - \frac{2D}{\alpha^2} t} \tag{92}$$

in the field of forces

$$F = \frac{2m_0 Dc}{\left(1 - \frac{2D}{\alpha^2} t\right)\left[\alpha^2 + \left(\frac{2D}{\alpha} t\right)^2\right]} \tag{93}$$

Considering that the type (92) changes are implied in gravitational interaction [61,62], it follows that the solidarity of the corpuscle movement with the movement of the associated progressive wave is accomplished by means of the appropriate gravitational field of the physical object; 3) The uniform movement V = c is obtained for null fractal force F = 0 and fractal speed U = 0, using condition x = ct. The fractal forces in the semi space. $-\infty \leq x \leq \bar{x}$ and $\bar{x} \leq x \leq +\infty$ are reciprocally compensated.

$$F\Big|_{-\infty}^{\bar{x}} = F\Big|_{\bar{x}}^{+\infty}$$

This means that the corpuscle in "free" motion simultaneously polarizes the "environment" of the wave behind x ≤ ct and in front of x ≥ ct, in such a manner that the resulting force has a symmetrical distribution as compared with the plane that contains the position of the noticeable object x = ct at any time moment t. Under such circumstances, the physical object uniformly moves (the corpuscle is located in the field of the associated wave).

12. THE CORPUSCLE AS A HOROLOGE AND ITS INCORPORATION IN THE ASSOCIATED WAVE. IMPLICATIONS

According to de Broglie's theory, the corpuscle must be associated to a horologe having the frequency equal to that of the associated progressive wave. Mathematically we can describe such an oscillator through the differential equation

$$\ddot{q}+\omega^2 q=0 \tag{94}$$

where ω defines the natural frequency of the oscillator as it is dictated by the environment (the wave), and the point above the symbol refers to the differential as compared with time. The most general solution of Equation (94) generally depends not on two arbitrary constants, as it is usually considered, but on three: the initial relevant coordinate, the initial speed and the phase of the harmonic oscillatory within the ensemble that structurally represents the environment (the isolated oscillator is an abstraction). Such a solution gives the relevant co-ordinate

$$q(t)=he^{i(\omega t+\Phi)}+\bar{h}e^{-i(\omega t+\Phi)} \tag{95}$$

where $\bar{h}$ refers to the complex conjugate of h and Φ is an initial phase specific to the individual movement of the oscillator. Such a notation allows us to solve a problem that we could name "the oscillators with the same frequency", such as Planck's resonators' ensemble—the basis of the quantum theory arguments in their old shape. That is, given an ensemble of oscillators having the same frequency in a space region, which is the relation between them?

The mathematical answer to this problem can be obtained if we note that what we want here is to find a mean to pass from a triplet of numbers—the initial conditions—of an oscillator towards the same triplet of another oscillator with the same frequency. This process (passing) implies a simple transitive continuous group with three parameters that can be built using a certain definition of the frequency.

We start from the idea that the ratio of two fundamental solutions of Equation (94) is a solution of Schwarts' non linear equation [63-65]

$$\frac{\mathrm{d}}{\mathrm{d}t}\left(\frac{\ddot{\tau}_0}{\dot{\tau}_0}\right)-\frac{1}{2}\left(\frac{\ddot{\tau}_0}{\dot{\tau}_0}\right)^2=2\omega^2,\quad \tau_0(t)\equiv e^{-2i\omega t} \tag{96}$$

This equation proves to be a veritable definition of frequency as a general characteristic of an ensemble of oscillators that can be scanned through a continuous group of three parameters. Indeed Equation (96) is invariant to the change of the dependent variable

$$\tau(t)=\frac{a\tau_0(t)+b}{c\tau_0(t)+\mathrm{d}} \tag{97}$$

which can be verified through direct calculation. Thus, $\tau(t)$ characterizes another oscillator with the same frequency which allows us to say that, starting from a standard oscillator we can scan the whole ensemble of oscillators of the same frequency when we let loose the three ratios a:b:c:d in Equation (97). We can make a more precise correspondence between a homographic change and an oscillator, by means of associating to each oscillator a personal $\tau(t)$ through equation

$$\tau_1(t)=\frac{h+\bar{h}k\tau_0(t)}{1+k\tau_0(t)}\qquad k\equiv e^{-2i\Phi} \tag{98}$$

Let us notice that τ_0, τ_1 can be freely used one instead the other, which leads to the next group of changes for the initial conditions

$$h'\rightarrow\frac{ah+b}{ch+\mathrm{d}} \tag{99a}$$

$$\bar{h}'\rightarrow\frac{a\bar{h}+b}{c\bar{h}+\mathrm{d}} \tag{99b}$$

$$k' \to k \cdot \frac{c\bar{h} + \mathrm{d}}{ch + \mathrm{d}} \tag{99c}$$

$$a, b, c, d \in R \tag{99d}$$

This is a simple transitive group: one and only one change of the group (the Barbilian group [63,64]) corresponds to a given set of values (a/c, b/c, d/c).

This group admits the 1-differential forms, absolutely invariant through the group [63]

$$\omega_0 = i\left(\frac{\mathrm{d}k}{k} - \frac{\mathrm{d}h + \mathrm{d}\bar{h}}{h - \bar{h}}\right), \omega_1 = \bar{\omega}_2 = \frac{\mathrm{d}h}{k\left(h - \bar{h}\right)} \tag{100}$$

and the 2-differential form

$$\frac{\mathrm{d}s^2}{\alpha^2} = \omega_0{}^2 - 4\omega_1\omega_2 = -\left(\frac{\mathrm{d}k}{k} - \frac{\mathrm{d}h - \mathrm{d}\bar{h}}{h - \bar{h}}\right)^2 + 4\frac{\mathrm{d}h\mathrm{d}\bar{h}}{\left(h - \bar{h}\right)^2},$$

$$\alpha = \text{const.} \tag{101}$$

respectively.

If we restrict the definition of a parallelism of directions in Levi-Civita manner [63]

$$\mathrm{d}\varphi = -\frac{\mathrm{d}u}{v} \tag{102}$$

With

$$h = u + iv, \bar{h} = u - iv, \; k = \mathrm{e}^{-i\varphi} \tag{103}$$

Barbilian's group invariates the metrics of Lobacevski's plane [63],

$$\frac{ds^2}{\alpha^2} = -\frac{du^2 + dv^2}{v^2} \tag{104}$$

Metrics (104) coincides with the differential invariant that is built with the complex scalar field of the speed,

$$\begin{aligned}\frac{ds^2}{\alpha^2} &\equiv d\phi d\bar{\phi} = (2D\mathrm{d}s - iD\mathrm{d}\ln\rho)(2D\mathrm{d}s + iD\mathrm{d}\ln\rho) \\ &= 4D^2(\mathrm{d}s)^2 + D^2\left(\frac{\mathrm{d}\rho}{\rho}\right)^2\end{aligned} \tag{105}$$

which admits the identities

$$\alpha = D, \tag{106a}$$

$$2\mathrm{d}s \equiv \mathrm{d}\quad \Phi = -\frac{\mathrm{d}u}{v}, \tag{106b}$$

$$\mathrm{d}\ln\rho \equiv \mathrm{d}\ln v \tag{106c}$$

Now, through a Matzner-Misner type principle one can obtain Ernst's principle of generating the symmetrical axial metrics [61,62]

$$\delta\int \frac{\nabla h \nabla \bar{h}}{\left(h - \bar{h}\right)^2}\gamma^{1/2}\mathrm{d}^3x = 0 \tag{106d}$$

where g = det g_{ab} with g_{ab} the metrics of the "environment".

Therefore, the incorporation of the corpuscle in the wave, considering that it functions as a horologe with the same frequency as that of the associated progressive wave, implies gravitation through Einstein's vacuum equations (equivalent to Ernst's principle (106d)). On the contrary, when the frequencies do not coincide, there is an induction of Stoler's group from the theory of coherent states (the parameter of the change is the very ratio of frequencies when creation and annihilation operators refer to a harmonic oscillator [63]

Let us note that the homographic changes (99) generalize the result (92). Moreover, if $a, b, c, d \in \mathbf{Z}$ then the Ernst type equations describe super-gravitation N = 1 [65].

13. CONCLUSIONS

Finally we can display the conclusions of this chapter as follows:

- A critical analysis of the hydrodynamic model of Madelung and of the double solution theory of de Broglie's theory of double solution was performed—departing from here, we built a fractal approximation of motion;
- We got the equation of motion of the physical object in the fractal approximation and the Eulerian case was studied;
- The flowing regimes of a rotational fractal fluid were studied;
- We studied the irrotational regime of a fractal fluid and the incorporation of the particle into the associated wave by generating a Schrödinger equation;
- The extended hydrodynamic model of scale relativity was built and the role of the fractal potential in the process of incorporation of the particle into the wave, specified;
- We indicated the mechanisms of wave-particle duality by their in phase coherences;
- We studied the wave-particle duality by stationary flow regimes of a fractal fluid which is coherent in phase, and by non-stationary flow regimes of an incoherent fractal fluid by means of a "polarization" type mechanism;
- Considering the particle as a singularity in the wave, we showed that its incorporation into the associated wave resulted in Einstein's

equations in vacuum—contrary, its non-incorporation led to the second quantification;

ACKNOWLEDGEMENTS

This paper was supported by the project POSDRU/88/1.5/ S/47646 of The European Social Fund and the project PERFORM-ERA "Postdoctoral Performance for Integration in the European Research Area" (ID-57649), financed by the European Social Fund and the Romanian Government.

REFERENCES

1. L. Nottale and J. Schneider, "Fractals and Non-Standard Analysis," Journal of Mathematical Physics, Vol. 25, No. 12, 1984, pp. 96-300.

2. L. Nottale, "Fractals and the Quantum Theory of SpaceTime," International Journal of Modern Physics, Vol. A4, No. 50, 1989, pp. 47-117.

3. L. Nottalle, "Fractal Space-Time and Microphysics: Towards a Theory of Scale Relativity," World Scientific, Singapore, 1993.

4. L. Nottale, "Scale Relativity and Fractal Space-Time: Applications to Quantum Physics, Cosmology and Chaotic Systems," Chaos, Solitons & Fractals, Vol. 7, No. 6, 1996, pp. 877-938.

5. L. Nottale, "The Theory of Scale Relativity: Non-Differentiable Geometry, Fractal Space-Time and Quantum Mechanics," Computing Anticipatory Systems: CASYS' 03- 33 Sixth International Conference, Liège, 11-16 August 2003, pp. 68-95.

6. J. Chaline, L. Nottale and P. Grou, "The Trees of Evolution: Universe S," Life Companies Societies, Paris, 2000.

7. J. Chaline, L. Nottale and P. Grou, "Flowers for Scrodinger Sscale Relativity and Its Applications," Editure Ellipses Marketing, Paris, 2009.

8. Harabagiu and M. Agop, "Hydrodyamic Model of Scale Relativity Theory," Bulletin of the Polytechnic Institute of Iasi, Vol. LV, No. 3-4, 2005, pp. 77-82.

9. Harabagiu, O. Niculescu, M. Colotin, T. D. Bibere, I. Gottlieb and M. Agop, "Particle in a Box by Means of Fractal Hydrodynamic Model," Romanian Reports in Physics, Vol. 61, No. 3, 2009, pp. 395-400.

10. Harabagiu, D. Magop and M. Agop, "Fractality and Quantum Mechanics," Ars Longa Publishing House, Iaşi, 2010.

11. M. Agop, L. Chicoş, P. Nica and A. Harabagiu, "Euler's Fluids and Non-Differentiable Space-Time," Far East Journal of Dynamical Systems, Vol. 10, No. 1, 2008, pp. 93-106.

12. M. Agop, A. Harabagiu and P. Nica, "Wave-Particle Duality through a Hydrodynamic Model of the Fractal Space Time Theory," Acta Physica Polonica A, Vol. 113, No. 6, 2008, pp. 1557-1574.

13. S. Ţiţeica, "Quantum Mechanics," Academiei Publishing House, Bucureşti, 1984.

14. Felsager, "Geometry, Particles and Fields," Odense University Press, Odense, 1981.

15. Peres, "Quantum Theory: Concepts and Methods," Kluwer Academic Publishers, Boston, 1993.

16. J. J. Sakurai and T. S. Fu, "Modern Quantum Mechanics," Addison-Wesley, Boston, 1994.

17. F. Halbwacs, "Theorie Relativiste Des Fluids a Spin," Gauthier-Villars, Paris, 1960.

18. R. Madelung, "A Hydrodynamic Interpretation of Quantum Mechanics," Journal for Physics, Vol. 40, No. 3-4, 1927, pp. 322-326.

19. Bohm and B. J. Hiley, "The Undivided Universe: An Ontological Interpretation of Quantum Theory," Routledge and Kegan Paul, London, 1993.

20. Dürr, S. Goldstein and N. Zanghi, "Quantum Mechanics, Randomness and Deterministic Reality," Physics Letters A, Vol. 172, No. 1-2, 1992, pp. 6-12.

21. Dürr, S. Goldstein and N. Zanchi, "A Global Equilibrium as the Foundation of Quantum Randomness," Foundations of Physics, Vol. 23, 1993, pp. 712-738.

22. D. Z. Albert, "Bohm's Alternative to Quantum Mechanics", Scientific American, Vol. 270, No. 2, 1994, pp. 32- 39.

23. J. S. Bell, "Speakable and Unspeakable in Quantum Mechanics," Cambridge University Press, Cambridge, 1987.

24. K. Berndl, D. Dürr, S. Goldstein, G. Peruzzi and N. Zanchi, "Existance of Trajectories for Bohmian Mechanics," International Journal of Theoretical Physics, Vol. 32, No. 12, 1993, pp. 2245-2251.

25. K. Berndl, D. Dürr, S. Goldstein, N. Zanchi, "Selfadjointness and the Existance of Deterministic Trajectories in Quantum Theory," In: M. Fannes, C. Maes and A. Verbeure, Eds., On Three Levels: Micro-, Meso-, and Macroscopic Approches in Physics, (NATO ASI Series B: Physics), Plenum, New York, 1994, pp. 429-434.

26. P. R. Holland, "The Quantum Theory of Motion," Cambridge University Press, Cambridge, 1993.

27. L. de Broglie, "Causal Interpretation Attempt and NonLinear Wave Mechanics: The Theory of Double Solution," Gauthier-Villars, Paris, 1956.

28. L. de Broglie, "The Measure Theory in Wave Mechanics," Gauthier-Villars, Paris, 1957.

29. L. de Broglie, "Wave Mechanics Interpretation," Journal of Physics Radium, Vol. 20, No. 12, 1959, pp. 963-979.

30. L. de Broglie, “Critical Study of the Current Foundations Interpretation of Wave Mechanics,” Gauthier-Villars, Paris, 1963.

31. L. de Broglie, “Thermodynamics of Isolated Particle (Hidden Thermodynamics of Particles),” Gauthier-Villars, Paris, 1964.

32. L. de Broglie, “Certainity and Uncertainity in Science,” Politică Publishing House, Bucureşti, 1980.

33. D. Bohm, “A Suggested Interpretation of Quantum Theory in Terms of ‘Hidden’ Variables I,” Physical Review, Vol. 85, No. 2, 1952, pp. 166-179.

34. D. Bohm, “A Suggested Interpretation of Quantum Theory in Terms of ‘Hidden Variables’: Part II,” Physical Review, Vol. 85, No. 2, 1952, pp. 180-193.

35. D. Bohm, “Proof that Probability Density Approaches $|\psi|$ in Causal Interpretation of Quantum Theory,” Physical Review, Vol. 89, No. 2, 1953, pp. 458-466.

36. L. Budei, “Fractal Models: Applications in Enviroment Arhitecture,” Publishing House University, Gh. Asachi, Iaşi, 2000.

37. M. Barnsley, “Fractals Everywhere: The First Course in Deterministic Fractal Geometry,” Academic Press, Boston, 1988.

38. Le Mehante, “Les Geometries Fractales,” Hermes, Paris, 1990.

39. Heck and J. M. Perdang, “Applying Fractals in Astronomy,” Springer Verlag, Berlin, 1991.

40. J. Feder and A. Aharony, “Fractals in Physics,” NorthHolland, Amsterdam, 1990.

41. P. Berge, Y. Pomeau and C. Vidal, “The Order in Chaos,” Wiley, New York, 1984.

42. J. F. Gouyet, “Physical Structures and Fractals,” Masson, Paris, 1992.

43. M. S. El Naschie, O. E. Rössler and I. Prigogine, "Quantum Mechanics, Diffusion and Chaotic Fractals", Elsevier, Oxford, 1995.

44. P. Weibel, G. Ord and O. E. Rössler, "Space Time Physics and Fractality: Festschroft in Honer of Mohamad El Naschie Vienna," Springer, New York, 2005.

45. Nelson, "Quantum Fluctuations," Princeton University Press, Princeton, 1985.

46. M. Agop, M. Colotin and V. Păun, "Haoticity, Fractality and Fields: Elements of Fractal Theory," Publishing House ArsLonga, Iaşi, 2009, pp. 12-46.

47. V. Chioroiu, L. Munteanu, P. Ştiucă and S. Donescu, "Introduction in Nanomechanics," Publishing House Academiei Române, Bucureşti, 2005.

48. D. K. Ferry and S. M. Goodnick, "Transport in Nanostructures," Cambridge University Press, Cambridge, 2001.

49. Y. Imry, "Introduction to Mesoscopic Physics," Oxford University Press, Oxford, 2002.

50. D. Benoit, "Physics of Semiconductor Microcavities from Fundamentals to Nano-Scale Decretes," Wiley-VCH Verlag GmbH, Weinheim, 2007.

51. Ciuti and I. Camsotto, "Quantum Fluids Effects and Parametric Instabilities in Microcavities," Physica Status Solidi B, Vol. 242, No. 11, 2005, p. 2224.

52. P. Poole, K. A. Farach and R. Creswick, "Super-Conductivity," Academic Press, San Diego, 1995.

53. Grössing, "Diffusion Waves in Sub-Quantum Thermodynamics: Resolution of Einstein's 'Particle-in-a-Box' Objection," 2008.

54. Mandelis, L. Nicolaides and Y. Chen, "Structure and the Reflectionless/Refractionless Nature of Parabolic Diffusion-Wave Fields, " Physical Review Letters, Vol. 87, No. 2, 2001, 4 pages.

55. Mandelis, "Diffusion Waves and Their Uses," Physics Today, Vol. 53, No. 8, 2000, p. 29.

56. S. Popescu, "Actual Issues in Self-Structuring Systems," Publishing House Tehnopress, Iasi, 2004.

57. R. Bittner, "Quantum Tunneling Dynamics Using Hydrodynamic Trajectories," Journal of Chemical Physics, Vol. 112, No. 22, 2000, p. 9703.

58. Ruelle and F. Takens, "On the Nature of Turbulence," Communications in Mathematical Physics, Vol. 20, No. 23, 1971, pp. 167-192,

59. D. Ruelle, "Strange Attractors," The Mathematical Intelligencer, Vol. 2, No. 3, 1979, pp. 126-137.

60. V. Munceleanu, D. Magop, C. Marin and M. Agop, "Fractal Models in Polymer Physics," Ars Longa Publishing House, Iasi, 2010.

61. J. Ernst, "New Formulation of the Axially Symemetric Gravitational Fielf Problem I," Physical Reviews, Vol. 167, No. 5, 1968, pp. 1175-1178.

62. J. Ernst, "Exterior Algebraic Derivation of Einstein Field Equation Employing a Generalized Basis," Journal of Mathematical Physics, Vol. 12, No. 11, 1971, pp. 2395-2398.

63. M. Agop and N. Mazilu, "Fundaments of Modern Physics," Junimea Publishing House, Iasi, 1989.

64. M. Agop and N. Mazilu, "The Crossroads of the Theories between Newton and Einstein Barbilian's Univers," Ars Longa, Iaşi, 2010.

65. M. B. Green, J. H. Schwarz and E. Witten, "Superstring Theory," Cambridge University Press, Cambridge, 1998.

3

CALCULATION OF THE APPROXIMATE ENERGY OF GROUND AND EXCITED STATIONARY STATES IN QUANTUM MECHANICS USING DELTA METHOD

Farrin Payandeh, Touraj Mohammadpour

Department of Physics, Payame Noor University (PNU), Tehran, Iran

ABSTRACT

In this paper, pursuing a new advised method called Delta method which is basically similar to variational method, we find the ground and excited states, according to a typical quantum Hamiltonian. Moreover, applying this method, the upper bound values for the eigenenergies of the so- called ground and excited states are estimated. We will show that this new method, is as beneficial as the traditional variational method which is common in deriving eigenenergies of some of the quantum Hamiltonians. This method helps physics students to broaden their knowledge about the possible mathematical ways; they can use to obtain eigenenergies of some quantum Hamiltonians. The advantage of Delta method to variational method is in its simplicity and reduction of the calculation procedures.

KEYWORDS

Quantum Mechanics, Eigenenergies, Alternative Methods, Delta Method

1. INTRODUCTION

Most problems encountered in quantum mechanics cannot be solved exactly. Exact solutions of the Schrodinger equation exist only for a few idealized systems. So, in order to solve general problems, one must resort to approximation methods. Up to now, a variety of such methods have been developed, and each has its own area of applicability. There are many methods for solving Schrodinger equation, i.e. perturbation theory [1] , the variational method [1] , and the WKB method [1] , Supersymmetry quantum mechanics [2] - [6] , Nikivorov-Uvarov method [7] - [9] , Romanovski polynomials in quantum mechanics [10] - [12] , etc. [12] - [23] .

Three conventional approximation methods for studying the stationary states corresponding to time-indepen- dent Hamiltonians, are: perturbation theory, the variational method, and the WKB method. Perturbation theory is based on the assumption that the problem we wish to solve is, in some sense, only slightly different from a problem that can be solved exactly. In the case where the deviation between the two problems is small, perturbation theory is suitable for calculating the contribution associated with this deviation; this contribution is then added as a correction to the energy and the wave function of the exactly solvable Hamiltonians. So perturbation theory builds on the known exact solutions to obtain approximate solutions.

But, about those systems whose Hamiltonians cannot be reduced to an exactly solvable part plus a small correction, the variational method or the WKB approximation are considered. The variational method is particularly useful in estimating the energy eigenvalues of the ground state and the first few excited states of a system for which one has only a qualitative idea about the form of the wave function. The WKB method is useful for finding the energy eigenvalues and wave functions of systems for which the classical limit is valid. Unlike perturbation theory, the variational and WKB methods do not require the existence of a closely related Hamiltonian that can be solved exactly [1] .

The application of the approximation methods to the study of stationary states consists of finding the energy eigenvalues E_n and the eigenfunctions $|\psi_n\rangle$ of a time-independent Hamiltonian $\hat{H}$ that does not have exact solutions:

$$\hat{H}|\psi_n\rangle = E_n|\psi_n\rangle$$

Depending on the structure of $\hat{H}$, we can use any of the three methods mentioned above to find the approximate solutions to this eigenvalue problem.

In this paper, we will use a new approximation method called Delta method for finding the ground and excited energy state of stationary states. This method, with a difference in the way of calculation, is somehow similar to the variational method at the beginning. Like the variational method, we first find the time-independent Hamiltonian $\hat{H}$ that does not have exact solution, using a supposed trial function and calculating the value of energy E_n in terms of a parameter α. Then, we will estimate the upper bound values for the eigenenergies of ground and excited states applying the Delta conditions $\Delta \geq 0$ and $D \leq 0$ on the second and third order equations, respectively. The difference between Delta and variational methods is that in variational method the derivative of E to α is calculated and the solution for α is obtained with affecting the conditions on E. But, in Delta method, we obtain the physical solutions with writing α in terms of E in the form of a two or three order equation and applying the mathematical Delta conditions. Moreover, in variational method, after calculating the derivative of E to α and equating it with zero to find the α which minimizes $E(\alpha)$, i.e. in Hydrogen atom, sometimes one of the solutions for α is infinite which is an unphysical solution, since it leads to a zero value for the ground or excited state energy. While, in delta method, only the physical solution is attained.

It should be noted that Delta method could be applied to all of the problems to be solved through variational method and exactly give the same result. However, the advantage of Delta method is first in its simplicity, and then in reduction of the calculation procedures.

In Section 2, we will have a review on the variational method [1] . In Section 3, we will explain Delta method and in Section 4, we will show the applicability and simplicity of Delta method with some examples.

2. VARIATIONAL METHOD

There exist systems whose Hamiltonians are known, but they cannot be solved exactly or by a perturbative treatment. That is, there is no closely related Hamiltonian that can be solved exactly or approximately by perturbation theory because the first order is not sufficiently accurate. One of the approximation methods that are suitable for solving such problems is the variational method, which is also called the Rayleigh-Ritz method. This method does not require knowledge of simpler Hamiltonians that can be solved exactly. The variational method is useful for determining upper bound values for the eigenenergies of a system whose Hamiltonian is known whereas its eigenvalues and eigenstates are not known. It is particularly useful for determining the ground state. It becomes quite cumbersome to determine the energy levels of the excited states.

In the context of the variational method, one does not attempt to solve the eigenvalue problem:

$$\hat{H}|\psi_n\rangle = E_n|\psi_n\rangle \tag{1}$$

But rather one uses a variational scheme to find the approximate eigenenergies and eigenfunctions from the variational equation:

$$\delta E(\psi) = 0 \tag{2}$$

where $E(\psi)$ is the expectation value of the energy in the state $|\psi\rangle$:

$$E(\psi)=\frac{\langle\psi|\hat{H}|\psi\rangle}{\langle\psi|\psi\rangle} \tag{3}$$

If $|\psi\rangle$ depends on a parameter α, $E(\psi)$ will also depend on α. The variational ansatz (2) enables us to vary α so as to minimize $E(\psi)$. The minimum value of $E(\psi)$ provides an upper limit approximation for the true energy of the system. The variational method is particularly useful for determining the ground state energy and its eigenstate without explicitly solving the Schrodinger equation. Not that for any (arbitrary) trial function $|\psi\rangle$ we choose, the energy E as given by (3) is always larger than the exact energy E_0:

$$E=\frac{\langle\psi|\hat{H}|\psi\rangle}{\langle\psi|\psi\rangle}\geq E_0 \tag{4}$$

The equality condition occurs only when $|\psi\rangle$ is proportional to the true ground state $|\psi_0\rangle$. Because, expanding the trial function $|\psi\rangle$ in terms of the exact eigenstates of $\hat{H}$:

$$|\psi\rangle=\sum_n a_n|\phi_n\rangle \tag{5}$$

With

$$\hat{H}|\phi_n\rangle=E_n|\phi_n\rangle \tag{6}$$

and since $E_0 \geq E_n$ for nondegenerate one-dimensional bound systems, we have:

$$E=\frac{\langle\psi|\hat{H}|\psi\rangle}{\langle\psi|\psi\rangle}=\frac{\sum_n |a_n|^2 E_n}{\sum_n |a_n|^2}\geq\frac{E_0\sum_n |a_n|^2}{\sum_n |a_n|^2}=E_0 \tag{7}$$

which proves (4).

To calculate the ground state energy, we need to carry out the following four steps:

- First, based on physical intuition, make an educated guess of a trial function that takes into account all the physical properties of the ground state (symmetries, number of nodes, smoothness, behavior at infinity, etc.). For the properties we are not sure about, they can be included in the trial function adjustable parameters $\alpha_1, \alpha_2, \cdots$ (i.e., $|\psi_0\rangle = |\psi_0(\alpha_1, \alpha_2, \cdots)\rangle$) which will account for the various possibilities of these unknown properties.

- Second, using (3), calculate the energy; this yields an expression which depends on the parameters $\alpha_1, \alpha_2, \cdots$:

$$E_0(\alpha_1, \alpha_2, \cdots) = \frac{\langle \psi_0(\alpha_1, \alpha_2, \cdots) | \hat{H} | \psi_0(\alpha_1, \alpha_2, \cdots) \rangle}{\langle \psi_0(\alpha_1, \alpha_2, \cdots) | \psi_0(\alpha_1, \alpha_2, \cdots) \rangle} \tag{8}$$

In most cases $|\psi_0(\alpha_1, \alpha_2, \cdots)\rangle$ will be assumed to be normalized; hence the denominator of this expression is equal to 1.

- Third, using (8) search for the minimum of $E_0(\alpha_1, \alpha_2, \cdots)$ by varying the adjustable parameters α_i until E_0 is minimized. That is, minimize $E(\alpha_1, \alpha_2, \cdots)$ with respect to $\alpha_1, \alpha_2, \cdots$:

$$\frac{\partial E_0(\alpha_1, \alpha_2, \cdots)}{\partial \alpha_i} = \frac{\partial}{\partial \alpha_i} \frac{\langle \psi_0(\alpha_1, \alpha_2, \cdots) | \hat{H} | \psi_0(\alpha_1, \alpha_2, \cdots) \rangle}{\langle \psi_0(\alpha_1, \alpha_2, \cdots) | \psi_0(\alpha_1, \alpha_2, \cdots) \rangle} = 0 \tag{9}$$

with $i = 1, 2, \cdots$. This gives the values of $(\alpha_{1_0}, \alpha_{2_0}, \cdots)$ that minimize E_0.

- Fourth, substitute these values of $(\alpha_{1_0}, \alpha_{2_0}, \cdots)$ into (8) to obtain the approximate value of the energy. The value of $E_0(\alpha_{1_0}, \alpha_{2_0}, \cdots)$ thus obtained, provides an upper bound for the exact ground state energy E_0. The exact ground state eigenstate $|\phi_0\rangle$ will then be approximated by the state $|\psi_0(\alpha_{1_0},\alpha_{2_0},\cdots)\rangle$.

About the energies of the excited states, it should be said that the variational method can also be used to find the approximate values for the energies of the first few exited states. For instance, to find the energy and eigenstate of the first excited state that will approximate E_1 and $|\phi_1\rangle$, we need to choose a trial function $|\psi_1\rangle$ that must be orthogonal to $|\psi_0\rangle$:

$$\langle\psi_1|\phi_0\rangle = 0 \tag{10}$$

Then proceed as we did in the case of the ground state. That is, solve the variational Equation (2) for $|\psi_1\rangle$:

$$\frac{\partial}{\partial\alpha_i}\frac{\langle\psi_1(\alpha_1,\alpha_2,\cdots)|\hat{H}|\psi_1(\alpha_1,\alpha_2,\cdots)\rangle}{\langle\psi_1(\alpha_1,\alpha_2,\cdots)|\psi_1(\alpha_1,\alpha_2,\cdots)\rangle} = 0; \quad (i = 1,2,\cdots) \tag{11}$$

Similarly, to evaluate the second excited state, we solve (2) for $|\psi_2\rangle$ and take into account the following two conditions:

$$\langle\psi_2|\psi_0\rangle = 0, \qquad \langle\psi_2|\psi_1\rangle = 0 \tag{12}$$

These conditions can be included in the variational problem by means of Lagrange multipliers, that is, by means of a constrained variational principle.

In this way, we can in principle evaluate any other excited state. However, the variational procedure becomes increasingly complicated as we deal with higher excited states. As a result, the method is mainly used to determine the ground state.

3. DELTA METHOD

In this section, we obtain the expectation value of energy in a supposed state $|\psi\rangle$, like the variational method, $|\psi\rangle$ depends on a parameter α, and finally $E(\psi)$ will depend on α, too. Then, instead of minimizing $E(\psi)$ leading to a minimum value of $E(\psi)$ as an upper approximate limit for the real energy of system, we use a new method in which we write the expectation value of energy in a supposed state $|\psi\rangle$ in the form of a second order equation in terms of α, i.e. $a\alpha^2 + b\alpha + c = 0$, and using the method of finding the solution of a second order equation, we calculate $\Delta = \sqrt{b^2 - 4ac}$ and then, write α in terms of $E(\psi)$, so that we have:

$$\alpha_{1,2} = \frac{-b \pm \sqrt{b^2 - 4ac}}{2a} \qquad (13)$$

where a,b,c are constants and the term $b^2 - 4ac$ will be a function of E. For having a solution for α in terms of E, we should have:

$$\Delta \geq 0 \Rightarrow b^2 - 4ac \geq 0 \qquad (14)$$

Then, using this inequality, just like the variational method, an upper limit for energy is obtained that leads to finding the approximate value of ground or excited state energy.

In some of the problems, the expectation value of energy in terms of α leads to a third order equation in terms of E in the form of $a\alpha^3 + b\alpha^2 + c\alpha + d = 0$ using which by writing the required conditions for having real roots of α in terms of E, one can find the desired solutions, i.e. the ground or excited state energy. Writing the required conditions for solving third order equation $ax^3 + bx^2 + cx + d = 0$, we first write the parameters f, q, D as the following:

$$P = \frac{1}{3}\left(\frac{3c}{a} - \frac{b^2}{a^2}\right) \qquad (15)$$

$$q=\frac{1}{27}\left(\frac{2b^3}{a^3}-\frac{9bc}{a^2}+\frac{27d}{a}\right) \tag{16}$$

$$D=\left(\frac{p}{3}\right)^3+\left(\frac{q}{2}\right)^2 \tag{17}$$

Now, according the sign of D, the solutions for the third order equation are obtained as:

$$\text{if } D\leq 0:\phi=\arccos\left[-\frac{q}{2}\left(\frac{|p|}{3}\right)^{-\frac{3}{2}}\right]$$

$$y_1=2\left(\frac{|p|}{3}\right)^{\frac{1}{2}}\cos\frac{\phi}{3} \tag{18}$$

$$y_{2,3}=-2\left(\frac{|p|}{3}\right)^{\frac{1}{2}}\cos\frac{\phi\pm\pi}{3} \tag{19}$$

Then, three distinct real solutions are obtained for α in terms of E. Moreover:

$$\text{if } D>0;\quad u=\left(-\frac{q}{2}+D^{\frac{1}{2}}\right)^{\frac{1}{3}},\quad v=\left(-\frac{q}{2}-D^{\frac{1}{2}}\right)^{\frac{1}{3}}$$

$$y_1=u+v, \tag{20}$$

$$y_{2,3}=-\frac{(u+v)}{2}\pm i\left(\frac{u-v}{2}\right)3^{\frac{1}{2}} \tag{21}$$

then, one real solution and two complex solutions are obtained [24] . Of course, since in physical problems α is a real scale factor, so without the need for calculating α in terms of E, it is sufficient to apply the Delta conditions $\Delta \geq 0$ and $D \leq 0$ on the second and third order equations respectively, and estimate the ground or excited state energies. It should be noted that the reason for using the condition $D \leq 0$ is to ensure the real solutions. This method can be applied for higher order equations in terms of α, too.

In the next section, we will consider a few examples of quantum mechanics and estimate their upper limit of the ground or excited state energies, using Delta method. We remark that this method can be applied as another approach to approximate solutions for all the problems of quantum mechanics to be solved through the variational method. In the case of selecting a suitable trial function, Delta method can be applied to a complex problem such as Poschl-Teller, Rosen-Morse, or another shape invariant potential in quantum mechanics, too. However, the main point is that proposing and guessing the suitable trial function which is based on physical intuition, is very important. For example, it can be shown that with some proposal trial wave functions and applying some simplification, exactly the same result could be achieved for Poschl-Teller and Rosen-Morse potentials through both variational and Delta methods. Of course, it should be noted that up to now, a variety of approximation methods have been developed, and each has its own area of applicability. However, the main purpose of introducing Delta method is to show that this new method, is as beneficial as the traditional variational method which is common in deriving eigenenergies of some of the quantum Hamiltonians, and to help physics students to broaden their knowledge about the possible mathematical ways to obtain eigenenergies of some quantum Hamiltonians with a simpler method than the conventional variational method.

4. EXAMPLES

In this section, using Delta method, we estimate the upper limit of the ground or excited state energies for a few examples of quantum mechanics that are obtained from the variational method, before.

Exp. 1

Calculating the ground and excited state energies of a one-dimensional harmonic oscillator using Delta method.

Choosing the trial function for ground state in the form of $\psi_0(x,\alpha) = Ae^{-\alpha x^2}$ and writing the expectation value of E_0 in terms of α:

$$E_0(\alpha) = \langle \psi_0 | H | \psi_0 \rangle = A^2 \int_{-\infty}^{+\infty} e^{-\alpha x^2} \left(-\frac{\hbar^2}{2m} \frac{d^2}{dx^2} + \frac{1}{2} m\omega^2 x^2 \right) e^{-\alpha x^2} dx$$

Neglecting the calculations, we get:

$$E_0(\alpha) = \frac{\hbar^2}{2m}\alpha + \frac{m\omega^2}{8\alpha}$$

In Delta method there is no need for obtaining the derivative of $E_0(\alpha)$ to α. It is just sufficient to form the second order equation for α in terms of E:

$$E_0(\alpha) = \frac{\hbar^2 4\alpha^2 + m^2\omega^2}{8m\alpha}$$

Using the Delta method, the solutions for α will be:

$$\alpha_{1,2} = \frac{8mE_0(\alpha) \pm \sqrt{(-8mE_0(\alpha))^2 - 4\alpha 4\hbar^2 m^2 \omega^2}}{2\alpha 4\hbar^2}$$

$$(8mE_0(\alpha))^2 \geq 16\hbar^2 m^2 \omega^2 \Rightarrow 8mE_0(\alpha) \geq 4\hbar m\omega \Rightarrow E_0(\alpha) \geq \frac{\hbar\omega}{2}$$

As it can be seen, $\frac{\hbar\omega}{2}$ is an upper limit for $E_0(\alpha)$ which is the value of the ground state energy. For obtaining the excited state energy, writing the trial function in the form of $\psi_1(x,\alpha) = Bxe^{-E_0(\alpha)x^2}$, we have:

$$E_1(\alpha)=\langle\psi_1(\alpha)|H|\psi_1(\alpha)\rangle=B^2\int_{-\infty}^{+\infty}xe^{-\alpha x^2}\left[-\frac{\hbar^2}{2m}\frac{d^2}{dx^2}+\frac{1}{2}m\omega^2x^2\right]xe^{-\alpha x^2}dx$$
$$=\frac{3\hbar^2}{2m}\alpha+\frac{3m\omega^2}{8\alpha}$$

Again, forming the second order equation of α in terms of $E_1(\alpha)$, we get:

$$12\hbar^2\alpha^2-8m\alpha E_1(\alpha)+3m^2\omega^2=0$$

and then,

$$\alpha_{1,2}=\frac{8mE_1(\alpha)\pm\sqrt{(-8mE_1(\alpha))^2-4\times12\hbar^2\times3m^2\omega^2}}{2\times12\hbar^2}$$

$$8mE_1(\alpha)\geq12\hbar m\omega\Rightarrow E_1(\alpha)\geq\frac{3\hbar\omega}{2}$$

which is the same value obtained from the variational method.

Exp. 2

Estimating the ground state energy of the Hydrogen atom using Delta method.

Choosing the trial function in the form of $\psi(r,\theta,\varphi)=e^{-\frac{r}{\alpha}}$, where α is a scale factor, energy is obtained as:

$$E(\alpha)=\frac{\langle\psi|H|\psi\rangle}{\langle\psi|\psi\rangle}=\frac{-\left\langle\psi\left|\frac{\hbar^2}{2m}+\frac{e^2}{r}\right|\psi\right\rangle}{\langle\psi|\psi\rangle}$$

$$\langle\psi|\psi\rangle=\int_0^{+\infty}r^2e^{-\frac{2r}{\alpha}}dr\int_0^{\pi}\sin\theta d\theta\int_0^{2\pi}d\varphi=\pi\alpha^3$$

$$-\left\langle\psi\left|\frac{e^2}{r}\right|\psi\right\rangle=-4\pi e^2\int_0^{+\infty}re^{-\frac{2r}{\alpha}}dr=-\pi e^2\alpha^2$$

$$-\left\langle \psi \left| \frac{\hbar^2}{2m} \nabla^2 \right| \psi \right\rangle = \frac{4\pi}{\alpha^2} \frac{\hbar^2}{2m} \int_0^{+\infty} r^2 e^{-\frac{2r}{\alpha}} dr = \frac{\hbar^2 \pi}{2m} \alpha$$

Now, using Delta method we have:

$$2m\alpha^2 E(\alpha) + 2m\alpha e^2 - \hbar^2 = 0 \rightarrow 8mE(\alpha)\hbar^2 \geq -\left(2me^2\right)^2$$

$$E(\alpha) \geq -\frac{me^4}{2\hbar^2}$$

that is the same estimated value for the ground state energy of the Hydrogen atom obtained from the variational method.

We should note that if we find $E(\alpha)$ for Hydrogen atom through variational method with the same trial function $\psi(r,\theta,\varphi) = e^{-\frac{r}{\alpha}}$:

$$E(\alpha) = \frac{\langle \psi | H | \psi \rangle}{\langle \psi | \psi \rangle} \rightarrow E(\alpha) = \frac{\hbar^2}{2m\alpha^2} - \frac{e^2}{\alpha} \rightarrow \frac{dE}{d\alpha} = 0 \rightarrow -\frac{\hbar^2}{m\alpha^3} + \frac{e^2}{\alpha} = 0$$

$$\rightarrow \frac{1}{\alpha^2}\left(-\frac{\hbar^2}{m\alpha} + e^2\right) = 0 \rightarrow \underbrace{E(\alpha) = -\frac{me^4}{2\hbar^2}}_{\alpha = \frac{\hbar^2}{me^2}}, \quad \underbrace{E(\alpha) = 0}_{\alpha = \infty}$$

we find that one of the solutions for α is infinite which is an unphysical solution, since it leads to a zero value for the ground or excited state energy. While, in Delta method, only the physical solution is attained.

Exp. 3

Estimation of the ground state energy of a one-dimensional harmonic oscillator by making use of Delta method with the following two trial functions:

(a) $\psi_0(x,\alpha) = Ae^{-\alpha|x|}$

(b) $\psi_0(x,\alpha) = \frac{A}{x^2 + \alpha}$

where α is a positive real number and A is the normalization constant.

(a) We have:

$$E(\alpha)=\frac{\left\langle \psi_0 \left| -\frac{\hbar^2}{2m}\frac{d^2}{dx^2}+\frac{1}{2}m\omega^2 x^2 \right| \psi_0 \right\rangle}{\langle \psi_0 | \psi_0 \rangle}$$

The final solution is:

$$2\hbar^2\alpha^4 + m^2\omega^2 = 4m\alpha^2 E(\alpha)$$

$$\alpha^2 = x \rightarrow 2\hbar^2\alpha^4 - 4m\alpha^2 E(\alpha) + m^2\omega^2 = 0$$

$$2\hbar^2 x^2 - 4mxE(\alpha) + m^2\omega^2 = 0$$

$$E(\alpha) \geq \frac{\sqrt{2}}{2}\hbar\omega$$

or

$$E(\alpha) \geq 0.707\hbar\omega,$$

which is the ground state energy of a one-dimensional harmonic oscillator with the above defined trial function and is the same result obtained from the variational method.

(b) We have:

$$E(\alpha)=\frac{\langle \psi_0 | H | \psi_0 \rangle}{\langle \psi_0 | \psi_0 \rangle}=\frac{\hbar^2}{4m\alpha}+\frac{1}{2}m\omega^2\alpha=\frac{\hbar^2+2m^2\omega^2\alpha^2}{4m\alpha}$$

$$\Rightarrow 2m^2\omega^2\alpha^2 + \hbar^2 = 4m\alpha E(\alpha)$$

therefore

$$E(\alpha) \geq \frac{\hbar\omega}{\sqrt{2}}$$

Again, it is quite in accordance with the resulted solution of variational method.

Exp. 4

Applying Delta method, we calculate the ground state energy for a particle of mass m which is bouncing vertically and elastically on a reflecting hard floor where $v(z) = \begin{cases} mgz & z > 0 \\ +\infty & z \leq 0 \end{cases}$ and g is the gravitational constant.

Choosing the trial function $\psi_0(z,x) = Aze^{-\alpha z}$, where α is a parameter and A is the normalization constant, we can show $A = z\alpha^{\frac{3}{2}}$ and therefore:

$$\psi_0(z,\alpha) = 2\sqrt{\alpha^3}ze^{-\alpha z}$$

$$E(\alpha) = \langle\psi_0|H|\psi_0\rangle = 4\alpha^3\int_0^{+\infty} ze^{-\alpha z}\left[-\frac{\hbar^2}{2m}\frac{d^2}{dz^2} + mgz\right]ze^{-\alpha z}dz$$

$$\hbar^2\alpha^3 + 3m^2g - 2m\alpha E(\alpha) = 0.$$

In the above third order equation of α in terms of $E(\alpha)$, using Equations (2) to (8) we have:

$$\begin{cases} p = \frac{1}{3}\left[\frac{3(-2mE)}{\hbar^2}\right] - \frac{0}{\hbar^2} = -\frac{2mE}{\hbar^2} \\ q = \frac{1}{27}\left(\frac{2(0)^3}{a^3} - \frac{2(0)c}{a^2} + \frac{(27)(3m^2g)}{\hbar^2}\right) = \frac{1}{27}\left((27)\frac{3m^2g}{\hbar^2}\right) = \frac{3m^2g}{\hbar^2} \\ D = \left(\frac{-2mE}{3\hbar^2}\right)^3 + \left(\frac{3m^2g}{2\hbar^2}\right)^2 \end{cases}$$

The third order equation has some defined solutions due to $D \leq 0$ or $D > 0$ or $D = 0$. The physical solution is obtained for $D \leq 0$ as:

$$-\left(\frac{2mE}{3\hbar^2}\right)^3 + \left(\frac{3m^2 g}{2\hbar^2}\right)^2 \leq 0$$

$$\frac{2mE}{3\hbar^2} \geq \left(\frac{3}{2}\right)^{\frac{2}{3}} \frac{m^{\frac{4}{3}} (g)^{\frac{2}{3}}}{\hbar^{\frac{4}{3}}}$$

$$E \geq \left(\frac{3}{2}\right)\left(\frac{9}{4}\right)^{\frac{1}{3}} \left(mg^2\hbar^2\right)^{\frac{1}{3}}$$

which is consistent with the solution obtained from variational method, i.e.

$$E \geq \left(\frac{3}{2}\right)\left(\frac{9}{2}\right)^{\frac{1}{3}} \left(\frac{1}{2}mg^2\hbar^2\right)^{\frac{1}{3}}.$$

Exp. 5

Estimation of the ground state energy and the corresponding wave function of a system consisting of two identical particles of spin $\frac{1}{2}$ that are confined to an isotropic three-dimensional harmonic oscillator potential of frequency ω, using delta method, when the two particles do not interact.

Choosing the trial function in the form of $\psi(\vec{r}_1, \vec{r}_2, \vec{s}_1, \vec{s}_c) = Ae^{-\alpha(r_1+r_2)}\psi$, we have:

$$H = -\frac{\hbar^2}{2m}\nabla_1^2 - \frac{\hbar^2}{2m}\nabla_2^2 + \frac{1}{2}m\omega^2 r_1^2 + \frac{1}{2}m\omega^2 r_2^2$$

$$1 = \langle\psi|\psi\rangle = |A|^2 \left[4\pi\int_0^{+\infty} e^{-2\alpha r_1} r_1^2 dr_1\right]\left[4\pi\int_0^{+\infty} e^{-2\alpha r_2} r_2^2 dr_2\right] = |A|^2 \frac{4\pi}{4\alpha^3}\frac{4\pi}{4\alpha^3} = \frac{|A|^2 \pi^2}{\alpha^6}$$

$$|A|^2 = \left(\frac{\alpha^3}{\pi}\right)^2 = \frac{\alpha^6}{\pi^2}$$

$$\langle\psi|H|\psi\rangle = 2|A|^2 \int_v e^{-2\alpha r_2} dv_2 \int_v e^{-\alpha r_1}\left[-\frac{\hbar^2}{2m}\nabla_1^2 + \frac{1}{2}m\omega^2 r_1^2\right]e^{-\alpha r_1} dv_1$$

$$= \frac{12m\omega^2}{\alpha^2} + \frac{4\hbar^2\alpha^2}{m} = E$$

Applying Delta method, we get:

$$4\hbar^2\alpha^4 - Em\alpha^2 + 12m^2\omega^2 = 0,$$

which with

$$\alpha^2 = x \rightarrow 4\hbar^2 x^2 - Emx^2 + 12m^2\omega^2 = 0$$

we will encounter the following condition:

$$E \geq 8\sqrt{3}\hbar\omega,$$

that is the same energy of the ground state obtained from the variational method.

5. CONCLUSION

In this paper, we used an alternative method of finding the ground and excited energies state of stationary states. Basically, it is the same as the variational method, where a time-independent Hamiltonian $\hat{H}$ with no exact solution is exploited, however, in this new method, we used a trial function and calculated the value of energy E_n in terms of a defined parameter α. Therefore, the estimation of the upper bound values and the excited state energies, is based on applying the Delta conditions $\Delta \geq 0$ and $D \leq 0$, which are included respectively, in appropriate second and third order equations. Presenting some examples, we showed that the corresponding results which have been obtained within this new method are in consistence with those derived from the traditional variational method. It is remarkable that all the problems to be solved through variational method, could be solved through Delta method and lead to exactly the same result with the difference that Delta method is simpler and the calculation procedures are less than that of the variational method.

REFERENCES

1. Zettili, N. (2009) Quantum Mechanics and Applications. John Wiley & Sons, Hoboken.

2. Domínguez-Hernández, S. and Fernández C, D.J. (2011) Rosen-Morse Potential and Its Supersymmetric Partners. International Journal of Theoretical Physics, 50, 1993-2001.

3. Díaz, J.I., Negro, J., Nieto, L.M. and Rosas-Ortiz, O. (1999) The Supersymmetric Modified Pöschl-Teller and Delta-Well Potentials. Journal of Physics A: Mathematical and General, 32, 8447-8460.

4. Negro, J., Nieto, L.M. and Rosas-Ortiz, O. (2000) Confluent Hypergeometric Equations and Related Solvable Potentials in Quantum Mechanics. Journal of Mathematical Physics, 41, 7964-7996.

5. Contreras-Astorga, A. and Fernández C, D.J. (2008) Supersymmetric Partners of the Trigonometric Pöschl-Teller Potentials. Journal of Physics A: Mathematical and General, 41, Article ID: 475303.

6. Compean, C.B. and Kirchbach, M. (2006) The Trigonometric Rosen-Morse Potential in Supersymmetric Quantum Mechanics and Its Exact Solutions. Journal of Physics A: Mathematical and General, 39, 547-557.

7. Karayer, H., Demirhan, D. and Büyükkilic, F. (2015) Extension of Nikivorov-Uvarov Method for the Solution of Heun Equation. Journal of Mathematical Physics, 56, 1-22.

8. Yasuk, F., Berkdemir, C. and Berkdemir, A. (2005) Exact Solutions of the Schrodinger Equation with Non-Central Potential by Nikiforov-Uvarov Method. Journal of Physics A: Mathematical and General, 38, Article No. 29.

9. Gönül, B. and Köksal, K. (2007) A Search on the Nikiforov-Uvarov Formalism. Physica Scripta, 75, 686-690.

10. Quesne, C. (2013) Extending Romanowski Polynomials in Quantum Mechanics. Journal of Mathematical Physics, 54, Article ID: 122103.

11. Weber, H.J. (2007) Connection between Real Polynomial Solutions of Hypergeometric-Type Differential Equations with Rodrigues Formula. Central European Journal of Mathematics, 5, 415-427.

12. Weber, H.J. (2007) Connection between Romanowski and Other Polynomials. Central European Journal of Mathematics, 5, 581-595.

13. Quesne, C. (2008) Exceptional Orthogonal Polynomials, Exactly Solvable Potentials and Supersymmetry. Journal of Physics A: Mathematical and Theoretical, 41, Article ID: 392001.

14. Quesne, C. (2009) Solvable Rational Potentials and Exceptional Orthogonal Polynomials in Supersymmetric Quantum Mechanics. SIGMA, 5, 1-24.

15. Quesne, C. (2011) Higher Order SUSY, Exactly Solvable Potentials, and Exceptional Orthogonal Polynomials. Modern Physics Letters A, 26, 1843-1852.

16. Quesne, C. (2011) Rationally-Extended Radial Oscillators and Laguerre Exceptional Orthogonal Polynomials in kth-Order SUSYQM. International Journal of Modern Physics A, 26, 5337-5347.

17. Quesne, C. (2012) Revisiting (Quasi-)Exactly Solvable Rational Extensions of the Morse Potential. International Journal of Modern Physics A, 27, Article ID: 1250073.

18. Quesne, C. (2012) Novel Enlarged Shape Invariance Property and Exactly Solvable Rational Extensions of the Rosen-Morse II and Eckart Potentials. SIGMA, 8, 1-19.

19. Odake, S. and Sasaki, R. (2009) Infinitely Many Shape Invariant Potentials and the New Orthogonal Polynomials. Physics Letters B, 679, 414-417.

20. Odake, S. and Sasaki, R. (2013) Extensions of Solvable Potentials with Finitely Many Discrete Eigenstates. Journal of Physics A: Mathematical and Theoretical, 46, Article ID: 235205.

21. Odake, S. and Sasaki, R. (2011) Exactly Solvable Quantum Mechanics and Infinite Families of Multi-Indexed Orthogonal Polynomials. Physics Letters B, 702, 164-170.

22. Grandati, Y. (2011) Solvable Rational Extensions of the Morse and Kepler-Coulomb Potentials. Journal of Mathematical Physics, 52, Article ID: 103505.

23. Ho, C.-L. (2011) Prepotential Approach to Solvable Rational Potentials and Exceptional Orthogonal Polynomials. Progress of Theoretical Physics, 126, 185-201

24. Woan, G. (2000) The Cambridge Handbook of Physics Formulas. Cambridge University Press, Cambridge, UK.

4

AVOIDING NEGATIVE PROBABILITIES IN QUANTUM MECHANICS

Golden Gadzirayi Nyambuya

Department of Applied Physics, National University of Science and Technology, Bulawayo, Republic of Zimbabwe

ABSTRACT

As currently understood since its discovery, the bare Klein-Gordon theory consists of negative quantum probabilities which are considered to be physically meaningless if not outright obsolete. Despite this annoying setback, these negative probabilities are what led the great Paul Dirac in 1928 to the esoteric discovery of the Dirac Equation. The Dirac Equation led to one of the greatest advances in our understanding of the physical world. In this reading, we ask the seemingly senseless question, "Do negative probabilities exist in quantum mechanics?" In an effort to answer this question, we arrive at the conclusion that depending on the choice one makes of the quantum probability current, one will obtain negative probabilities. We thus propose a new quantum probability current of the Klein-Gordon theory. This quantum probability current leads directly to positive definite quantum probabilities. Because these negative probabilities are in the bare Klein-Gordon theory, intrinsically a result of negative energies, the fact that we here arrive at a theory with positive probabilities, means that negative energy particles are not to be considered problematic as is the case in the bare Klein-Gordon theory. From an abstract—objective stand-point; in comparison with positive energy particles, the corollary is that negative energy particles should have equal chances to exist. As to why these negative energy particles do not exist, this is analogous to asking why is it that Dirac's antimatter does not exist in equal proportions with matter. This problem of why negative energy particles do not exist in equal proportions with positive energy particles is a problem that needs to be solved by a future theory.

KEYWORDS

Klein-Gordon Equation; Schrödinger Equation; Probability; Negative Probability

1. INTRODUCTION

"My work always tried to unite the Truth with the Beautiful, but when I had to choose one or the other, I usually chose the Beautiful."

—Hermann Klaus Hugo Weyl (1885-1955)

If one accepts the bare Klein-Gordon theory as it is currently understood since its discovery in 1927 by Oskar Klein (1894-1977, of Sweden) and Walter Gordon (1893-1939, of Germany), then, there is no doubt that they will accept without fail that negative quantum mechanical probabilities do exist in the bare Klein-Gordon theory. Solemnly, by a combination of a deep and rare curiosity, fortune, and serendipity, than by natural design, the existence of these negative probabilities in the Klein-Gordon theory is what led the eminent British physicist Paul Adrien Maurice Dirac (1902-1984) to his landmark discovery of the Dirac Equation [1,2]. Needless to say, but perhaps as a way of expressing our deepest admiration of this great achievement, the Dirac Equation ranks amongst the greatest and most noble equations of physics. Eighty four years on since its discovery (i.e., 1928-2012), the Dirac Equation is an equation whose wealth of knowledge cannot be said to have been completely deciphered and fathomed but is in the process thereof.

Like other deep-thinking physicists of his time, right from the-word-go, Dirac passionately objected to the notion of negative probabilities implied by the KleinGordon theory. This led him to silently embark on a noble scientific journey of the mind whose final destination was to successfully solve this persistent and nagging problem of negative probability. Dirac had hoped that by eliminating the negative probabilities, he would concurrently eliminate the negative energies—he hoped for the rare fortuity of hitting two birds with one stone. Alas, that did not happen. Only the possibility of negative probabilities vanished while the negative energies stubbornly reared their "ugly head" in Dirac's new vision. On completion of his seemingly divine scientific journey of the mind, he arrived at his esoteric equation, which accurately describes the Electron.

Why do we say Dirac silently embarked on his quest for the Dirac Equation? Well the answer to this is that— for example, during a (tea/lunch) break at the 1927 Solvay conference attended by the great Danish physicist Neils Henrik David Bohr (1885-1962), Dirac, and many other towering figures of the last half century in physics; Dirac was asked by Neils Bohr what he was working on, to which he replied: "I'm trying to take the square root of something..." meaning the square root of the KleinGordon Equation—in physics, this is strange because it meant taking the square root of an operator. Later, Dirac recalled that he continued on by saying he was trying to find a relativistic quantum theory of the Electron, to which Bohr commented, "But Klein has already solved that problem". Dirac then tried to explain he was not satisfied with the (Klein-Gordon) solution because it involved a 2nd order equation in the time and space derivatives. Dirac was simply not open; he was a man of notoriously very few words, he meant every word he said; in a nutshell, he was economic with his words.

Further, Dirac was a man of great mathematical subtleness; it is this quality which led him to the Dirac Equation. He believed that one must follow the mathematics to where it would lead and in so doing he unlocked a great wealth of ideas such as magnetic monopoles, the variation of the gravitational constant amongst others. On the same pedestal—in 1942, all in an effort to forge ahead to seek new frontiers in the field of physical knowledge; Dirac back-pedalled on negative probabilities when he delved once again into the then un-chartered waters of negative probability when he wrote a paper entitled: "The Physical Interpretation of Quantum Mechanics" where he introduced the concept of negative probabilities [3]. In the introductory section of his watershed reading, he said:

"Negative energies and probabilities should not be considered as [mere] nonsense. They are well-defined concepts mathematically, like a negative of money".

Fifty five years later after Dirac's musings, i.e. in 1987 toward the end of his fruitful life, another great mind, the flamboyant and charismatic American physicist, Richard Feynman (1918-1988), took the idea further when he argued that, no one objects to using negative numbers in calculations, although "minus three apples" is not a valid concept in real life [so, it should be reasonable to consider negative probabilities too]. Further into the shores of the unknown, he [Feynman] argued not only how negative probabilities could possibly be useful in probability calculations, but as well how probabilities above unity may be useful [4]. Really? What would a probability above unity really mean?

The ideas of Dirac and Feynman on negative probability have not gained much support. To ourselves, negative probabilities, even if they may be well defined mathematical concepts as Dirac and Feynman believe or want us to believe, they [negative probabilities] are physically meaningless and obsolete; they signify something "sinister" in the finer detail of the theory in question. We like to view these ideas of negative probability as nothing more than highlighting and dramatising the desperation by physicists to make sense of nonsense all in an effort to find a natural explanation of nonsense. We think and hold that nonsense is nonsense and should be left that way; one should simply let the sleeping dogs lay. We have to be ruthless in our dismissal of these negative probabilities; we can only hope that the reader will pardon us for this.

The root of negative probabilities and all that seems bizarre in the world of probability theory is the KleinGordon theory. If it could be shown that the Klein-Gordon theory is devoid of these, it would render Dirac and Feynman's effort worthless—and; order, harmony and tranquillity will certainly be restored in the world of probability theory.

This would mean the chapter of negative probabilities is closed altogether. The endeavour of this reading is to point out that the Klein-Gordon theory is devoid of these negative probabilities, hence possibly bringing to a complete standstill Dirac and Feynman's efforts.

If only physicists had extended the British-German physicist Max Born (1882-1970)'s idea that the magnitude of the wavefunction gives the probability density function [7]; that is, extend this idea so that it applies to all quantum mechanical wavefunctions, then, we would never have landed on these rough, bizarre and uncertain shores of negative probabilities. As will be argued, what physicists have done is to carry over the probability current density found in the Schrödinger theory directly into the Klein-Gordon theory, in which process the quantum probability of the Klein-Gordon theory is constrained in a manner that allows for negative probabilities. Our suggestion, if correct as we would like to believe, is that instead of carrying over the probability current density found in the Schrödinger theory into the Klein-Gordon theory, we need to do things the other way round, that is, we have to carry over the probability density function of the Schrödinger theory into the Klein-Gordon theory. Simple, we must generalize Born's idea, that is:

Born's idea that the wavefunction represents the probability amplitude and its magnitude represents the probability; this idea must be generalized so that it is applicable to any general wavefunction that purports to describe undulatory material particles.

In this way, we constrain the resultant probability current density of the Klein-Gordon theory so as to allow only for positive definite probabilities—this is where we believe the problem in the negative probabilities lies. In the end, we obtain a Klein-Gordon theory that is devoid of negative probabilities; this of cause leads us to an objective world since all the probabilities are not only positive but positive definite. Notice that Schrödinger's wavefunction together with Dirac's wavefunction all conform to Born's idea but the Klein-Gordon wavefunction does not. Why? We herein propose an answer to this question.

Now, to wind up this section, we shall give the synopsis of this reading; it is organised as follows: in the next section, we present the

Schrödinger quantum mechanical probability theory as it is understood in the present day. In § (3), we also present the Klein-Gordon probability theory as it is understood in the present day. In § (4), we go onto the main theme of the present reading, where we demonstrate that if one makes an appropriate choice of the Klein Gordon probability current, one obtains a Klein-Gordon theory that is free from negative probabilities. While § (2) and § (3) may seem trivial to the quantum mechanically erudite reader, it is worthwhile that we mention that we have taken the decision to go through these sections [i.e. § (2) and § (3)] for nothing other than instructive purposes. The well versed reader will obviously have to skip these and go straight to § (4). In § (5), we give the overall discussion and conclusions drawn thereof.

2. SCHRÖDINGER THEORY

While in search of the Schrödinger Equation the great Austrian physicist, Erwin Rudolf Josef Alexander Schrö- dinger (1887-1961) first arrived at the Klein-Gordon Equation but discarded it because it did not give the correct predictions for the hydrogen atom. Schrödinger, was largely motivated to successfully search for the Schrö- dinger Equation after a thoughtful remark by the eminent Professor, Peter Joseph William Debye (1884-1966, of Austria), at the end of a lecture that he [Schrödinger] delivered on de Broglie's waves at the University of Vienna where he [Schrödinger] was working.

Professor Debye who was the head of the physics research group, on hearing of the de Broglie waves, he asked Schrödinger to explain these to the rest of the research group. So, the great Schrödinger weighed up to the task. At the end of the lecture, Professor Debye remarked that it seemed childish to talk of waves without a corresponding wave equation?! This proved to be Schrödinger's great moment of inspiration that would immortalize his name in the annals of human history. What is the wave equation describing the de Broglie waves?

In his 1924 doctoral thesis, which was nearly turned down [thanks to the pre-eminent French physicist Paul Langevin (1872-1946)'s wisdom and Einstein's influence and stature[1]], the French Prince, Louis Victor Pierre Raymond de Broglie (1892-1987) hypothesised that there is a duality between waves and matter; he gave a formula for the matter waves which stated that the wavelength of material particles is inversely proportional to the momentum of the matter particle in question. However, in his proposal, he did not propose the corresponding wave equation for these matter waves. As Professor Debye pointed out to Schrödinger, logic dictates that every wave must be described by a corresponding wave equation. The deep-and-agile Schrödinger saw immediately the depth and breadth of Professor Debye's question and as the lore holds; he [Schrödinger] went into "hiding" for about six months in search of the Schrödinger Equation which he successfully found at the end of his esoteric sojourn which was not without tribulation and trials [6]. The equation he found is:

$$-\frac{\hbar^2}{2m}\nabla^2\Psi + V\Psi = i\hbar\frac{\partial\Psi}{\partial t}, \tag{1}$$

where the arcane symbol Ψ is the Schrödinger wavefunction (or probability amplitude), $\hbar$ is Planck's normalized constant, m is the mass of the particle in question and ∇ is the usual 3D differential operator and t is time. In presenting his equation in 1926, Schrödinger interpreted the magnitude of the wavefunction as giving the density of the Electronic charge of the atom. However, Max Born [3] gave a radically different interpretation, where this quantity [magnitude of the wavefunction] is assumed to represent the probability that the atom is in a given state. In this way, Born ushered physics into the depth and realm of probability calculus. To this day, physicists do not agree on how to interpret this wavefunction but the general consensus is that, it is a probability function.

2.1. Schrödinger Probability Current Density

For instructive purposes, we present here the usual way in which one arrives at the expression of the Schrödinger probability current density. To do this, we have to take the Schrödinger Equation, divide it throughout by $i\hbar$ and then multiply the resultant by the complex conjugate of the wavefunction, i.e.:

$$\Psi^* \frac{\partial \Psi}{\partial t} = \frac{i\hbar}{2m}\left(\Psi^* \nabla^2 \Psi - \frac{2imV}{\hbar}\Psi^* \Psi\right) \tag{2}$$

Further, taking the complex conjugate of this same equation and then multiplying it by the wavefunction, one arrives at:

$$\Psi \frac{\partial \Psi^*}{\partial t} = -\frac{i\hbar}{2m}\left(\Psi \nabla^2 \Psi^* - \frac{2imV}{\hbar}\Psi^* \Psi\right), \tag{3}$$

and now adding these two equations, one obtains:

$$\Psi^* \frac{\partial \Psi}{\partial t} + \Psi \frac{\partial \Psi^*}{\partial t} = \frac{i\hbar}{2m}\left(\Psi^* \nabla^2 \Psi - \Psi \nabla^2 \Psi^*\right). \tag{4}$$

If we set $\rho = \Psi^* \Psi$, which is the probability density function, then:

$$\frac{\partial \rho}{\partial t} = \Psi^* \frac{\partial \Psi}{\partial t} + \Psi \frac{\partial \Psi^*}{\partial t}. \tag{5}$$

For the right handside of Equation (4), we will have:

$$\begin{aligned} &\frac{i\hbar}{2m}\left(\Psi^* \nabla^2 \Psi - \Psi \nabla^2 \Psi^*\right) \\ &= -\nabla \cdot \left[\frac{i\hbar}{2m}\left(\Psi^* \nabla \Psi - \Psi \nabla \Psi^*\right)\right], \end{aligned} \tag{6}$$

and setting:

$$\boldsymbol{J}_\rho^S = \frac{i\hbar}{2m}\left(\Psi^* \nabla \Psi - \Psi \nabla \Psi^*\right), \tag{7}$$

which is the Schrödinger probability current density, then, Equation (4) reduces the probability conservation equation, i.e.:

$$\frac{\partial \rho}{\partial t} = -\nabla \cdot \boldsymbol{J}_{\rho}^{S}. \tag{8}$$

Now, this is not the only continuity equation that can be written out of the Schrödinger Equation. The quantity $(\rho, \boldsymbol{J}_{\rho}^{S})$ comprises the four probability current density. In the next section, we shall write another continuity equation out of the Schrödinger Equation not in terms of the probability density function but in terms of the probability amplitude, Ψ.

2.2. Schrödinger Equation as a Continuity Equation

To write down another continuity equation out of the Schrödinger Equation in terms of the probability amplitude, we know that, for any general smooth, square integrable and differentiable space and time varying function (or field) $\Psi = \Psi(\boldsymbol{r}, t)$, from the chain rule, the following holds true always:

$$\begin{aligned}\Psi &\equiv \int_{r_0}^{r} (\nabla\Psi)\cdot \mathrm{d}\boldsymbol{r} = \int_{t_0}^{t} (\nabla\Psi)\cdot\left(\frac{\mathrm{d}\boldsymbol{r}}{\mathrm{d}t}\right)\mathrm{d}t \\ &= \int_{t_0}^{t} (\boldsymbol{v}\cdot\nabla\Psi)\,\mathrm{d}t,\end{aligned} \tag{9}$$

where $\boldsymbol{v}$ is the guiding velocity of the particle described by the field (i.e., the function Ψ) and $\mathrm{d}\boldsymbol{r} = \mathrm{d}x\boldsymbol{i} + \mathrm{d}y\boldsymbol{j} + \mathrm{d}z\boldsymbol{k}$ is a differential element of the position vector, where the $\boldsymbol{i}$, $\boldsymbol{j}$, $\boldsymbol{k}'$s are the usual orthogonal unit vectors along the x, y and z-axis respectively. In the integral (9), the initial conditions are such that at time t_0, the particle is at position $\boldsymbol{r} = \boldsymbol{r}_0$. At $(\boldsymbol{r}, t) = (\boldsymbol{r}_0, t_0)$, the wavefunction Ψ is such that $\Psi(\boldsymbol{r}_0, t_0) = 0$. This initial condition $[i.e.\ \Psi(\boldsymbol{r}_0, t_0) = 0]$ is required by one of the postulates of QM where the wavefunction must vanish at the end points where the particle is confined[2]. The identity (9) is central to the main result of this reading—it is the guide and lodestar of the present thesis. If the reader has any objections to the final conclusions that we shall arrive at, then they must object to this identity forthwith.

Before we proceed, we need to clearly define $\boldsymbol{v}$. Our interpretation of is exactly the same as that handed down to us by the great theoretical physicist—AmericanBritish Professor, David Joseph Bohm (1927-1992), in his Pilot Wave Theory which is popularly known as Bohmian Mechanics[3][8,9]. Remember that all quantum mechanical wavefunctions are functions of the phase $S = \boldsymbol{k} \cdot \boldsymbol{r} \pm \omega t$ i.e. $\Psi = \Psi(\boldsymbol{k} \cdot \boldsymbol{r} \pm \omega t)$: where $\boldsymbol{k}$ and ω are the wave number and angular frequency of the particle. Since according to the de Broglie's wave-particle duality hypothesis $\boldsymbol{p} = \hbar \boldsymbol{k}$ and $E = \hbar\omega$ where $\boldsymbol{p}$ and E are the momentum and energy of the particle; it follows that $\Psi = \Psi(\boldsymbol{p} \cdot \boldsymbol{r} \pm Et)$ and $S = \boldsymbol{p} \cdot \boldsymbol{r} \pm Et$. As first noted by de Broglie4 (1927) and latter [8,9] (independently of de Broglie's 1927 work), the wavefunction can be written in the form $\Psi = Re^{iS/\hbar}$ where $R = R(\boldsymbol{r}, t)$ is a real function which acts as a modulation function of the particle wavepacket. With the wavefunction written in this form, the guiding velocity $\boldsymbol{v}$ of the particle then becomes:

$$\boldsymbol{v} = \left(\frac{\hbar}{m}\right) \mathrm{Im}\left(\frac{\Psi^* \nabla \Psi}{\Psi^* \Psi}\right), \tag{10}$$

where m is the mass of the particle under consideration [8,9, de Broglie,^{1927}]. In (10), the operator "Im" extracts only the imaginary part of a complex function or complex number i.e.: if $z = x + iy$ is a complex number, then $\mathrm{Im}(z) = y$ and if $\mathrm{Im}[F(x)] = H(x)$ is a complex valued function where (G, H) are real valued functions of x, then $\mathrm{Im}[F(x)] = H(x)$. Thus, for any smooth square integrable and differentiable function Ψ the particle guiding velocity is given by (10).

Now moving on; the identity (9) can further be simplified. Since $\nabla \cdot \boldsymbol{v} = 0$, it follows that $\boldsymbol{v} \cdot \nabla \Psi = \nabla \cdot (\boldsymbol{v}\Psi)$ because according to the chain rule:

$$\nabla \cdot (\boldsymbol{v}\Psi) = \boldsymbol{v} \cdot \nabla \Psi + \Psi \nabla \cdot \boldsymbol{v} = \boldsymbol{v} \cdot \nabla \Psi + 0 = \boldsymbol{v} \cdot \nabla \Psi.$$

With $\boldsymbol{v} \cdot \nabla \Psi = \nabla \cdot (\boldsymbol{v}\Psi)$ given, (9) becomes:

$$\Psi \equiv \nabla \cdot \left[\int_{t_0}^{t} (\boldsymbol{v}\Psi) \mathrm{d}t \right] = \nabla \cdot \left[\boldsymbol{v} \int_{t_0}^{t} \Psi \mathrm{d}t \right]. \tag{11}$$

All we want with the relation (9) is to have the operator "∇" outside of the integral operator "∫". This is important for latter purposes. In the far right of (11), the velocity $\boldsymbol{v}$ has been removed from inside the integral sign. This is possible because $\mathbf{v}$ has no explicit dependence on time. Any change in $\boldsymbol{v}$ with respect to time happens not explicitly but implicitly.

Now, as we are going to demonstrate; using the identity (11), one can recast after a few basic algebraic operations, the Schrödinger Equation (1) into a continuity equation. To achieve this, we begin by dividing (1) throughout by "$-i\hbar$", where we will have:

$$\frac{\partial \Psi}{\partial t} = -\frac{i\hbar}{2m} \nabla^2 \Psi + \frac{i}{\hbar} V \Psi. \tag{12}$$

Now, taking the first term on the right hand side of Equation (12), it is clear that:

$$-\frac{i\hbar}{2m} \nabla^2 \Psi = \nabla \cdot \left[-\frac{i\hbar}{2m} \nabla \Psi \right]. \tag{13}$$

For the second term on the right hand side of (12), using the identity (11) where $V\Psi$ replaces Ψ in this identity i.e. $V\Psi \mapsto \Psi$, it is clear from this identity that:

$$\frac{i}{\hbar} V \Psi = \nabla \cdot \left[\frac{i\boldsymbol{v}}{\hbar} \int_{t_0}^{t} V \Psi \mathrm{d}t \right]. \tag{14}$$

Now, adding (13) and (14) , we will have:

$$\frac{\partial \Psi}{\partial t} = -\frac{i\hbar}{2m}\nabla^2\Psi + \frac{i}{\hbar}V\Psi$$
$$= -\nabla \cdot \left[+ \frac{i\hbar}{2m}\nabla\Psi - \frac{i\mathrm{v}}{\hbar}\int_{t_0}^{t} V\Psi \mathrm{d}t \right]. \tag{15}$$

This can now be written as:

$$\frac{\partial \Psi}{\partial t} + \nabla \cdot \boldsymbol{J}_{\Psi}^{S} = 0, \tag{16}$$

where the new current term $\boldsymbol{J}_{\Psi}^{S}$ is given by:

$$\boldsymbol{J}_{\Psi}^{S} = \frac{i\hbar}{2m}\nabla\Psi - \frac{i\boldsymbol{v}}{\hbar}\int_{t_0}^{t} V\Psi \mathrm{d}t. \tag{17}$$

What this means is that there is a corresponding conserved probability amplitude current for the probability current. The quantity $J_{\mu}^{\Psi} = \left(\Psi, \boldsymbol{J}_{\Psi}^{S}\right) : (\mu = 0,1,2,3)$ comprises the four probability amplitude current. Though this is a very trivial result, if it is correct (as we believe) and acceptable, it is a new result in QM.

3. KLEIN-GORDON PROBABILITY AS CURRENTLY UNDERSTOOD

For a free particle of rest mass m_0 and wavefunction Ψ, the Klein-Gordon Equation describing this particle is given by:

$$\nabla^2\Psi - \frac{1}{c^2}\frac{\partial^2\Psi}{\partial t^2} = \left(\frac{m_0 c}{\hbar}\right)^2 \Psi. \tag{18}$$

This equation is named after the physicists Oskar Klein and Walter Gordon, who in 1927 proposed it as an equation describing relativistic

Electrons. The KleinGordon Equation was first considered as a quantum mechanical wave equation by Schrödinger in his search for an equation describing de Broglie waves. In his final presentation in January 1926 where he proposed the Schrödinger Equation, Schrödinger discarded this equation because when the Coulomb potential is in-cooperated for the case of an Electron-Proton system, it did not give the correct predictions for the hydrogen atom as Schrödinger expected.

Now, we would like to develop for the Klein-Gordon Equation the usual expressions for probability density function and the corresponding probability current density similar to the Schrödinger case. This is a task that is considered a bit tricky as compared to the Schrödinger case because the Klein-Gordon Equation is second differential equation. If we take the probability density function $\rho = \Psi^*\Psi$ and then differentiate it with respect to time, what we get is (5), and from this point, if we are to follow the Schrödinger prescription, we should be able to arrive at a continuity equation containing the probability density function and the corresponding probability current density by substituting the time derivatives of the wavefunction.

Now—here lies the problem; most textbooks will tell you that one is not able to proceed to find the continuity equation for the above equation from the Schrödinger prescription simply because the Klein-Gordon Equation does not have a first-order derivative that would enable a straight substitution [see e.g. Ref. 10, pp. 576-578]. So what is typically done is to work backwards, that is start from the known Schrödinger probability current density and proceed from there to see if one can find a corresponding probability density function. As we all know, one does arrive at a continuity equation, namely:

$$\frac{\partial \rho_{KG}}{\partial t} + \nabla \cdot \boldsymbol{J}_{\rho}^{S} = 0, \tag{19}$$

where the Klein-Gordon probability is given by:

$$\rho_{KG} = \frac{i\hbar}{2m_0c^2}\left(\Psi^* \frac{\partial \Psi}{\partial t} - \Psi \frac{\partial \Psi^*}{\partial t}\right). \tag{20}$$

There is no need for us to go through the full derivation of the Klein-Gordon probability continuity equation as this can readily be found in most textbooks of quantum mechanics. Further, there is no need to demonstrate that this probability density function leads to negative probabilities for particles of negative energy as this is well anchored in most quantum mechanics textbooks. What we shall do is to point out that there is a loophole in this derivation and this loophole is deeply embedded in the fact that:

The Klein-Gordon Equation is not a first order differential equation which would allow for a smooth and straight forward substitution of the time derivative of the magnitude of the Klein-Gordon wavefunction into Equation (4) directly from the Klein-Gordon Equation, so as to derive the probability continuity equation; because of this, one has to seek other alternative means.

Why not force the Klein-Gordon Equation to produce a probability current density under these conditions? The legitimate rules of mathematics allow for this, why not go for it?! This is our borne of contention.

It is perhaps important that we mention here that in 1934, two other distinguished Austrians, Wolfgang Pauli (1904-1982) and Victor Frederick Weisskopf (1908-2002) discovered what is hailed as a suitable interpretation of the Klein-Gordon Equation within the scope of quantum field theory. Treating it [Klein-Gordon wavefunction] like a field equation analogous to Maxwell's Equations for an electromagnetic field, they quantized it, so that Ψ became an operator [11]. This made the Klein-Gordon theory more acceptable and since then, there appears to have been some acceptable and believable order in the Klein-Gordon world.

As will be seen latter in §(4), if the main reason for adding the Klein-Gordon probability density function that leads to negative probabilities is that it emerges from the continuity equation constructed out of the second order differential Klein-Gordon Equation; then, this way of arriving at the probability density function can be challenged as there is another way to arrive at a continuity equation from the Klein-Gordon Equation, this equation involves the magnitude of the Klein-Gordon wavefunction. After all, no one has made a direct measurement to test the correctness or lack thereof the Klein-Gordon probability and the Klein-Gordon probability current density, it is just but an "agreed upon interpretation".

Klein-Gordon Probability Amplitude Current

First things first, we need to categorically state one thing which is clear to all: which is that, the Klein-Gordon Equation is not cast in stone; the meaning of which is that it can be rewritten in different but mathematically equivalent forms provided one applies permissible and legitimate mathematical operations to it. So, just as we have done in the Schrödinger case, we will write down the continuity equation of the Klein-Gordon Equation which involves the probability amplitude (Ψ) and not the probability density function (ρ). By integrating the Klein-Gordon Equation with respect to time throughout, one consequently recasts this equation into the form:

$$\frac{1}{c^2}\frac{\partial \Psi}{\partial t} - \nabla \cdot \left[\nabla\left(\int \Psi \mathrm{d}t\right)\right] = -\left(\frac{m_0 c}{\hbar}\right)^2 \int \Psi \mathrm{d}t. \tag{21}$$

In the above equation—for our convenience vis-a-vis the labour of typing, we have not put the limits to the integral as we have done in (11); from here and after, we shall no-longer put the limits. It shall be assumed that the reader knows them. That said, written in this form i.e. the Klein-Gordon Equation in the form (21), one can easily construct a continuity equation. The strategy here being championed is to rewrite the second order Klein-Gordon differential equation as a first order differential equation in terms of time as is the case with the Schrödinger Equation. Having done this; using the fact stated in (11), it is not difficult to deduce that (21) can further be written as:

$$\frac{\partial \Psi}{\partial t} = \nabla \cdot \left[\nabla\left(\int c^2 \Psi \mathrm{d}t\right) - \left(\frac{m_0 c^2}{\hbar}\right)^2 \left(\iint \Psi \mathrm{d}t \mathrm{d}t\right) \boldsymbol{v}\right]. \tag{22}$$

Thus setting:

$$\boldsymbol{J}_{\Psi}^{KG} = \nabla\left(\int c^2 \Psi \mathrm{d}t\right) - \left(\frac{m_0 c^2}{\hbar}\right)^2 \left(\int\int \Psi \mathrm{d}t \mathrm{d}t\right)\boldsymbol{v}, \tag{23}$$

it is easy to see that:

$$\frac{\partial \Psi}{\partial t} + \nabla \cdot \boldsymbol{J}_{\Psi}^{KG} = 0. \tag{24}$$

This is our desired equation. What this means is that the probability amplitude has a corresponding current. In the language of Einstein's Special Theory of Relativity (STR), it means we can talk of a four probability amplitude comprising the probability amplitude and the probability amplitude current i.e. $\left(\rho, \boldsymbol{J}_{\Psi}^{KG}\right)$.

4. NEW KLEIN-GORDON PROBABILITY CONTINUITY EQUATION

Now, we come to the main theme of this reading. In (21) we have written the Klein-Gordon Equation with the time derivative to first order. For convenience, we shall rewrite this Equation (21) here as:

$$\frac{\partial \Psi}{\partial t} = \nabla \cdot \left[\nabla\left(\int c^2 \Psi \mathrm{d}t\right)\right] - \left(\frac{m_0 c^2}{\hbar}\right)^2 \int \Psi \mathrm{d}t. \tag{25}$$

Now, multiplying this equation throughout by the complex conjugate of the wavefunction, that is:

$$\Psi^* \frac{\partial \Psi}{\partial t} = \Psi^* \nabla \cdot \left[\nabla\left(\int c^2 \Psi \mathrm{d}t\right)\right] - \left(\frac{m_0 c^2}{\hbar}\right)^2 \Psi^* \int \Psi \mathrm{d}t, \tag{26}$$

and then taking the complex conjugate of this same equation and then multiplying it by the wavefunction, one arrives at:

$$\Psi\frac{\partial\Psi^*}{\partial t}=\Psi\nabla\cdot\left[\nabla\left(\int c^2\Psi^*\mathrm{d}t\right)\right]$$
$$-\left(\frac{m_0c^2}{\hbar}\right)^2\Psi\int\Psi^*\mathrm{d}t. \tag{27}$$

Adding these two Equations (26) and (27), one obtains:

$$\Psi^*\frac{\partial\Psi}{\partial t}+\Psi\frac{\partial\Psi^*}{\partial t}=c^2\left(\Psi^*\int\nabla^2\Psi+\Psi\int\nabla^2\Psi^*\right)\mathrm{d}t$$
$$-\left(\frac{m_0c^2}{\hbar}\right)^2\left(\Psi^*\int\Psi+\Psi\int\Psi^*\right)\mathrm{d}t. \tag{28}$$

The left hand side is obviously equal to the time derivative of the probably density function $\rho=\Psi^*\Psi$ i.e. $\mathrm{d}\rho/\mathrm{d}t=\mathrm{d}\left(\Psi^*\Psi\right)/\mathrm{d}t$. To simplify the right hand side, we have to make use of the identity in Equation (11); doing so, we will have:

$$c^2\left(\Psi^*\int\nabla^2\Psi+\Psi\int\nabla^2\Psi^*\right)\mathrm{d}t$$
$$=-\nabla\cdot\left[-\boldsymbol{v}\int\left\{c^2\left(\Psi^*\int\nabla^2\Psi+\Psi\int\nabla^2\Psi^*\right)\mathrm{d}t\right\}\mathrm{d}t\right], \tag{29}$$

and upon further reduction, the above becomes:

$$-\left(\frac{m_0c^2}{\hbar}\right)^2\left(\Psi^*\int\Psi+\Psi\int\Psi^*\right)\mathrm{d}t$$
$$=-\nabla\cdot\left[\mathbf{v}\int\left\{\left(\frac{m_0c^2}{\hbar}\right)^2\left(\Psi^*\int\Psi+\Psi\int\Psi^*\right)\mathrm{d}t\right\}\mathrm{d}t\right]. \tag{30}$$

Now, adding (29) and (30), we obtain, $-\nabla\cdot\boldsymbol{J}^{KG}$, where $\boldsymbol{J}^{KG}$ is the new Klein-Gordon probability current density such that $\boldsymbol{J}^{KG}=\rho\boldsymbol{v}$ where ρ is what we shall call the Klein-Gordon probability charge density. This charge density is such that:

$$\rho = -\int \left[\int \left\{ c^2 \left(\Psi^* \int \nabla^2 \Psi + \Psi \int \nabla^2 \Psi^* \right) dt \right\} - \left\{ \left(\frac{m_0 c^2}{\hbar} \right)^2 \left(\Psi^* \int \Psi + \Psi \int \Psi^* \right) dt \right\} \right] dt. \tag{31}$$

This new Klein-Gordon probability current density $\boldsymbol{J}^{KG} = \rho \boldsymbol{v}$ is what leads us to positive definite probabilities5. All our seemingly naïve efforts lead us to recast the Klein-Gordon Equation into the new continuity equation:

$$\frac{\partial \rho}{\partial t} + \nabla \cdot \boldsymbol{J}_{\rho}^{KG} = 0. \tag{32}$$

In this manner, just as in the Schrödinger case, the magnitude $\left(\rho = \Psi^* \Psi \geq 0 \right)$ of the wavefunction gives a positive definite probability and is part of a four current. The quantity $\left(\rho, \boldsymbol{J}_{\rho}^{KG} \right)$ is the new four Klein-Gordon probability current that leads to positive definite probabilities via the new definition of the Klein-Gordon probability current $\boldsymbol{J}_{\rho}^{KG}$. We feel and believe this approach is the correct approach to understanding the Klein-Gordon Equation. It contains no negative probabilities but real and objective probabilities just as in the Schrödinger case. It is our modest and vested opinion that it is much easier to try and understand $\boldsymbol{J}_{\rho}^{KG}$ as the new Klein-Gordon probability current density, than to try and justify negative probabilities as Dirac, Feynman and many others have attempted (with great pain), only to fail.

5. DISCUSSION AND CONCLUSION

5.1. General Discussion

At this juncture, we are of the candid view that the reader will concur with us that—from the rather trivial presentation made herein; the existence of negative quantum mechanical probabilities depends on the choice of the probability current density that one has made. In moving from Schrödinger's theory (which officially was first to be discovered) to the Klein-Gordon theory, it is the probability current density that is held sacrosanct [e.g. Ref. 10, pp. 576-578], the meaning or suggestion of which is that it must be the important quantity, otherwise there really would be no reason to preserve it. We ask our dear reader; could there be any other better reason to do so besides that it is a "sacrosanct quantity"? We have suggested otherwise, that it is the probability density function in the Schrödinger theory that must be held sacrosanct when we move over to the Klein-Gordon theory. This way of looking at the Klein-Gordon theory resolves (once and for all-time) the negative probabilities that bedevil and bewilder this theory. Because we here have shown that these negative probabilities can be gotten rid of completely, we believe this reading is a significant contribution to physics insofar as the endeavours to deciphering and fathoming the meaning of Klein-Gordon's negative probabilities is concerned.

In all history of human experience, it is important to note that no single experiment has been performed to date to directly measure the probability and the probability current implied by the quantum mechanical wavefunction. For example, the wavefunction of the hydrogen atom as deduced from the Schrödinger Equation is known yet no one has measured directly that the Electron in the hydrogen atom is found at the position that it is expected to be with the predicted frequency. The probability interpretation is an interpretation that strongly appears to work very well, especially when dealing with ensembles. What this means is that the currently accepted Klein-Gordon probability can be revised as we have done. If what is required of this probability is that it satisfies the continuity equation where this probability density function has a corresponding probability current density —well then—we have shown that there exists such a positive definite probability satisfying the continuity equation. It is difficult to dismiss the present approach (proposal). Whether or not mainstream science will seriously consider our approach (proposal made herein), only time will tell.

If our suggestion is correct and acceptable (as we believe it to be), then, one is lead to wonder what trajectory physics might have taken if what we have just presented were known to Dirac and his contemporaries.

This is so especially given that Dirac was largely motivated by the desire to get reed of the negative probabilities that appear in the Klein-Gordon theory. There is nothing exotic or new about the ideas that we have presented, it is just a different way of looking at things. The only thing that appears to make this approach of importance, is that it allows us to settle once and for all-time the nagging problem of negative probabilities.

Clearly, if the present presentation was available and acceptable to Dirac before the advent of his equation; then, if he [Dirac] was to discover the Dirac Equation as he did, he would have arrived at it from a different point of departure altogether. I wonder what his motivation would have been. Trying to imagine what his point of departure would have been, leads me into the wilderness of thought—analogous to chasing after the rainbow, the wind or one's own shadow. Perhaps, out of mathematical curiosity, beauty and elegance, he simply would have sought for an equation linear in both the time and space derivatives.

On the other hand, since negative probabilities are intrinsically tied to negative energy-mass particles, the non-existence of negative probabilities would mean that the existence of negative energy-mass particles has no problem in principle. This invariably means that negative energy-mass particles must be considered without any prejudice whatsoever as they have equal legitimacy to exist. Our only concern would be what these negative energy-mass particles are; are they Dirac's antimatter, or Dirac's sea of invisible energy-mass particles? Current thinking due to Richard Feynman is that negative energy-mass particles are antiparticles. These antiparticles have positive energy and the reason for this is that they are thought to be negative energy-mass particles moving back in time, in which case they would appear to have positive energy-mass [see e.g. Ref. 12, pp. 66-70]. The perfect symmetry of Dirac's theory allows a negative energy-mass particle that is moving forward in time to look identical to a positive energy-mass particle moving back in time.

5.2. Conclusions

Assuming the correctness (or acceptableness) of the ideas presented herein, we hereby make the following conclusion:

1) This reading has demonstrated that negative probabilities can be avoided in quantum mechanics by making an appropriate choice of the Klein-Gordon probability current.

2) By avoiding the negative probabilities as suggested herein, it is seen that negative energy particles will still exist with the important difference that they are no longer attached to the meaningless negative probabilities as is the case in the original Klein-Gordon theory. These negative energy particles have a real positive definite probability of existence.

ACKNOWLEDGEMENTS

I am grateful to the four anonymous Reviewers for their effort that greatly improved and refined the arguments presented herein. Further, I am grateful to the National University of Science and Technology's Research & Innovation Department and Research Board for their unremitting support rendered toward my research endeavours; of particular mention, Dr. P. Makoni and Prof. Y. S. Naik's unwavering support. This publication proudly acknowledges a GRANT from the National University of Science and Technology's Research Board.

REFERENCES

1. P. A. M. Dirac, Proceedings of the Royal Society of London, Vol. 117, 1928, pp. 351-361.
2. P. A. M. Dirac, Proceedings of the Royal Society of London, Vol. 118, 1928, pp. 610-612

3. P. A. M. Dirac, Proceedings of the Royal Society of London, Vol. 180, 1942, pp. 1-39.

4. R. P. Feynman, "Negative Probability," In: D. Bohm, F. D. Peat and B. Hiley, Eds., Quantum Implications, Routledge & Kegan Paul Ltd., London, New York, 1987, pp. 235-248.

5. G. Ghirardi, "Sneaking a Look at God's Cards," Princeton University Press, Princeton, 2005.

6. E. Schrödinger, Physical Review, Vol. 28, 1926, pp. 1049- 1070.

7. M. Born. Zeitschrift für Physik, Vol. 37, 1926, pp. 863- 867.

8. D. Bohm, Physical Review, Vol. 89, 1953, pp. 458-466.

9. D. Bohm, Physical Review, Vol. 85, 1952, pp. 166-193.

10. B. R. Desai. "Quantum Mechanics with Basic Field Theory," Cambridge University Press, Cambridge, 2010.

11. W. Pauli and V. F. Weisskopf, Helvetica Physica Acta, Vol. 7, 1934, pp. 709-731.

12. R. P. Feynman, "Quantum Electrodynamics: Lecture Notes," W. A. Benjamin Inc., 1961.

NOTES

[1]Instead of rejecting de Broglie's doctoral thesis, the cautious Langevin decided that the greatest living scientist of the day—Albert Einstein—must be consulted on the matter before a fateful decision is reached. The agile Einstein immediately endorsed the idea with the remark that de Broglie had gone a step further from where he [Einstein] had left the issue of wave-particle duality. In his [Einstein] own words: "I believe it is a first feeble ray of light on this worst of our physics enigmas. I, too, have found something which speaks for his construction" [see e.g. Ref. 5, p. 21]. Behold! To a "sleepy world", unknowingly, like Russia's Sputnik flying past the United States of America in the cover of darkness, de

Broglie had just opened the scientific Pandora's box, a very rich scientific gold mine which to the present we are still trying to understand.

[2]For example, in one dimension, for a particle confined in the region $(0 \leq x \leq a)$, we have $\Psi(0,t) = \Psi(a,t) = 0$.

[3]This theory is also called the de Broglie-Bohm theory, the Pilot-Wave model, or the Causal Interpretation of Quantum Mechanics.

[4]de Broglie, L., 1928, in Solvay 928, p. 119: Rapports et Discussions du Cinqui ème Conseil de Physique tenu à Bruxelles du 24 au 29 Octobre 1927 sous les Auspices de l'Institut International de Physique Solvay, Paris: Gauthier-Villars.

[5]In comparison—the expression for the Schrödindger probability current density 'is more beautiful' than the new expression for the Klein-Gordon probability current. In-accordance with Weyl's doctrine, our work is always to unite the Truth with the Beautiful. In all honesty, there is no derivable Truth in negative probabilities of Klein and Gordon, thus how can we unite these together into an everlasting ornament? There is obviously an element of truth in the new positive definite Klein-Gordon probability. Somehow, there must exist a hidden beauty in the new Klein-Gordon probability current.

5

COMPLEX MATTER SPACE AND RELATIVISTIC QUANTUM MECHANICS

Reza Ahangar

Department of Mathematic, Texas A & M University-Kingsville, Kingsville, TX, USA

ABSTRACT

The Special Relativity Theory cannot recognize speed faster than light. New assumption will be imposed that matter has two intrinsic components, 1) mass, and 2) charge, that is $M = m + iq$. The mass will be measured by real number system and charged by an imaginary unit. This article presents a Complex Matter Space in Relativistic Quantum Mechanics. We are hoping that this approach will help us to present a general view of energy and momentum in Complex Matter Space. The conclusion of this article on Complex Matter Space (CMS) theory will lead help to a better understanding toward the conversion of mass and energy equation, unifying the forces, and unifying relativity and quantum mechanics.

KEYWORDS

Complex Matter, Complex Mass, Complex Energy and Momentum

1. INTRODUCTION TO THE COMPLEX MATTER SPACE

It is common in mathematics to define an abstract space like metric, Hilbert, or Banach spaces without physical or experimental verifications. In fact logical approach with definitions, some related axioms, and presentation of some examples may be sufficient to accept and apply to practical life problems. But for constructing a theoretical physical space, in addition to the logical and theoretical approach, an experimental or physical justification is needed (see [1] and [2]).

In 1928, the British Physicist Paul Dirac realized that when electrons are moving very close to the speed of light, special relativity needs to be incorporated which lead to particles with negative mass (see [3]). The existence of a particle moving faster than light was presented by many authors including, Sudarshan who discussed the physics of complex mass (see [4]).

During the past decades, various theoretical approaches were presented to unify relativity and quantum mechanics which include complex space-time and axiomatic quantum field theory (see [5] and [6]).

The content of the paper on "Imaginary Relativity" is a right direction toward Complex Matter Space (CMS) but not sufficient for the unifying forces. The CMS postulates may answer some of the questions in this article's concluding remarks. This could be a reason for the statement that "the author does not know where the imaginary world is located or what its nature is?" (see [7]).

We will present postulates that a particle has two components that can be demonstrated in a complex plane as a real component mass (m) and imaginary component charge (q):

$$M = m + iq \tag{1.1}$$

We will show that these assumptions lead us to unify the electromagnetic force and gravitational force.

For example the interaction force between two particles M and M' can be described by

$$F = k\frac{M \cdot M'}{r^2} = k\frac{(m+iq)\cdot(m'+iq')}{r^2} \tag{1.2}$$

We can observe that, after expansion of (1.2), two terms on the right hand side represent gravitational forces and electromagnetic forces. But two other imaginary components seemed to me not to represent any forces.

Assume that the terminology "matter" in general is an undefined term. In this paper, in addition to the postulates of Special Relativity Theory (SRT), we accept the following new postulates:

Complex Matter Space (CMS)	
Postulate (I):	Postulate (II):
Mass and charge are two intrinsic components of the matter and they can be measured and determined in a complex plane. The real part is equal to the mass and the imaginary part is the charge of the particle. That is $M = m + iq$.	In a one dimensional space, position and velocity are assumed to be real numbers and the rate of change of the position of a particle can be assumed to be faster than the speed of light.

The position x(t) and the velocity dx/dt will stay a real number. The speed of light remains a constant real number, c. By this agreement, and contrary to the paper "Imaginary Relativity" by Rezaei, we are not accepting variation of the speed of light or considering it as a complex number. For decades, the discovery of the nature of other forces was a dream. Acceptance of this postulate leads to the necessity to reaffirm the entire physics, Newtonian mechanics, relativity, and quantum mechanics.

The organization of this paper is the following: A new view of complex matter will be presented based on two postulates. We use a geometric interpretation of complex matter to justify new postulates in a complex coordinate plane.

By these new postulates of Mass and Energy we can restructure the relativistic view of complex matter, the Einstein equation of mass-energy. Finally, energy and momentum in Relativistic Quantum Mechanics will be recalculated in the Complex Matter Space.

2. GEOMETRIC VIEW OF THE COMPLEX MATTER USING IMAGINARY OBJECT

Traveling Beyond the Speed of Light: According to the Special Theory of Relativity, we may not be allowed to cross the boundary of the speed of light c in the real world. But if we can pretend that we exist, in the imaginary world then it may be allowed to go faster than light, since our conjugate space will be the real world.

Assume that we are traveling by a rocket which is taking off from the earth to go cruising in space. How much energy is required to accelerate to the speed of light? It would be impossible for a particle to accelerate to a velocity of light c because the speed of light is constant in a vacuum for all inertial observers. What do we observe?

1) Observers in all inertial frames on the earth would measure the same value for c and in fact*infinite energies* are required to accelerate objects to the speed of light.

2) If we can break the wall of speed of light and travel faster than light, an observer would see things going *crazy* around—by SRT time dilation the time would freeze around us—mass will approach infinity.

Experiment of Moving the Image of a Light Faster than the Speed of Light: Assume a source of a light located at the epicenter of two spheres (O, c) and (O, R) where OM = R ? c whose radius R is much greater than the speed of light c. This sphere can be real but radius can be astronomically large. To perform this experiment, one can select c = 1 as a unit of the measurement thus R ? 1.

The source of light at the center (O) emitting the light signal in each direction and rotating with the rate of $\omega(t)=\frac{d\theta}{dt}$.

The light is traveling at the speed of the c and it will take seconds to show the image on the dome. We will measure the linear speed of the image on the dome.

For simplicity two circles of radius c = 1 and R with the images of mapping of A and B on the larger circle is M and N, where their arc-lengths are s and s' respectively.

In Figure 1 the light signal emits from the center (O) and transfers the image of the point A to a point M on the second circle, while the point is moving on the circumference of the first circle. In the following simple approach, we will demonstrate that the image point A is moving faster than the point M on the circumference of the second circle.

We denote T(A) = M and similarly T(B) = N such that length of the arc AB is s and T(s) = S.

$$s=r\cdot\theta=c\cdot\theta\Rightarrow\frac{ds}{dt}=\frac{ds}{d\theta}\cdot\frac{d\theta}{dt}=c\frac{d\theta}{dt}$$

Let us assume that the measure of the angle (AOB) = theta varies with the rate of a fixed real number, for example k = 2, therefore $\frac{d\theta}{dt}=2$. Substituting in the previous relation implies that

$$s=r\cdot\theta=c\cdot\theta\Rightarrow\frac{ds}{dt}=c\frac{d\theta}{dt}=2c$$

Thus, the image M which is moving on the larger circle (O, R) with the linear speed of 2c is faster than light.

Complex Coordinate in Minkowski Space-Time: Absolute time may be an undefined word and we try to postulate, define, and measure it by its changes, variations, and motions with other physical quantities in space, i.e. physical motion of particles, rotation of the earth around the sun, or clock's hands on a circle. The unit of time practically is what we perceive about how much the physical objects change position, velocity, energy, or mass. How do we define one hour, one minute, or one second? They are the measurements of the physical changes that we try to demonstrate by arc-length on a clock using the unit of distance or radians on a circle.

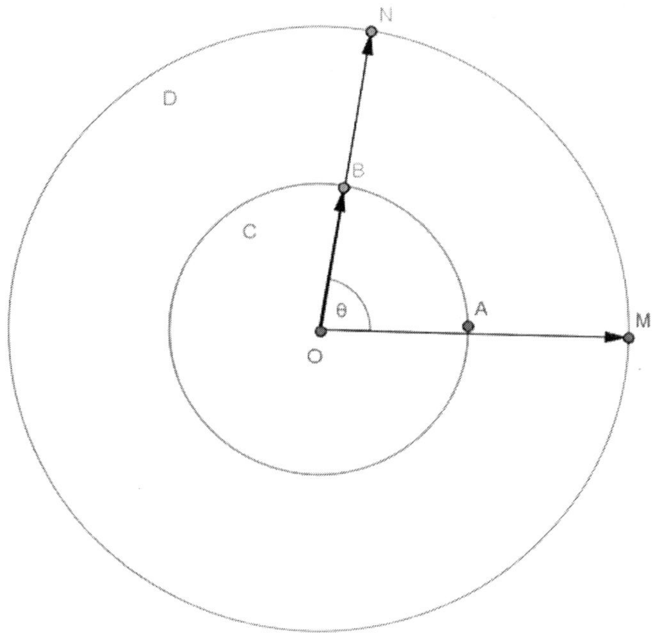

Figure 1. The light signal is emitting from the center (O) and transfers the image of the point A to the second circle M while the point is moving on the circumference of the first circle and the image is moving faster on the circumference of the second circle.

Are they real measurements of time? In the following mathematical argument we introduce $i = \sqrt{-1}$ as a unit of time on the imaginary axis. Let us consider a vector $\psi = (u, v, w, t)$ such that

$$\langle \psi_1, \psi_2 \rangle = u_1 u_2 + v_1 v_2 + w_2 w_2 + (it_1)(it_2)$$
$$\langle \psi_1, \psi_2 \rangle = u_1 u_2 + v_1 v_2 + w_2 w_2 - (t_1)(t_2) \tag{2.1}$$

Magnitude of a Vector 4D-Complex Matter Space: We may use the norm in the inner product space as a distance or magnitude of a vector $\psi = (u, v, w, t)$

$$\langle \psi, \psi \rangle = \|\psi\|^2 = u^2 + v^2 + w^2 - t^2$$

Verifying the CMS Postulate (II): According to time and length dilation in SRT, the value of time and length variation

$$\Delta t = \frac{\Delta t'}{\sqrt{1-\left(\frac{v}{c}\right)^2}} = \gamma \cdot \Delta t' \text{ and } \Delta L = \frac{\Delta L'}{\sqrt{1-\left(\frac{v}{c}\right)^2}} = \gamma \cdot \Delta L' \tag{2.2}$$

such that

(a) $\gamma = \dfrac{1}{\sqrt{1-\left(\frac{v}{c}\right)^2}} = \dfrac{1}{i\sqrt{\left(\frac{v}{c}\right)^2 - 1}} = \dfrac{-i}{\sqrt{\left(\frac{v}{c}\right)^2 - 1}}$ and (b) $\gamma^* = \dfrac{i}{\sqrt{\left(\frac{v}{c}\right)^2 - 1}}$ (2.3)

can be interpreted for the velocity v as a real number greater than the speed of light.

This is one indication that time may have an imaginary unit.

Assuming a complex time variation

$t_1 = a + ib$ and

The relation (2.2) can be expressed by

$$\Delta t = \frac{p + iq}{\sqrt{1-\left(\frac{v}{c}\right)^2}} = \gamma \cdot (p + iq)$$

Case (I): the motion velocity v < c.

The values of α , β , and γ will remain real numbers. Consequently, according to (2.2) velocity v and distance L will be real numbers.

$$v(t) = \lim_{\Delta t \to 0} \frac{\Delta L}{\Delta t} = \lim_{\Delta t \to 0} \frac{\gamma \cdot \Delta L'}{\gamma \cdot \Delta t'} = \lim_{\Delta t \to 0} \frac{\Delta L'}{\Delta t'}$$

Case (II): the motion velocity v > c: If the particle velocity is faster than the speed of light (v > c) then both γ and γ^* are pure imaginary numbers and using (2.2) the following shows that

$$v(t) = \lim_{\Delta t \to 0} \frac{\Delta L}{\Delta t} = \lim_{\Delta t \to 0} \frac{\gamma \cdot \Delta L'}{\gamma \cdot \Delta t'} = \lim_{\Delta t \to 0} \frac{\Delta L'}{\Delta t'}$$

is a real number. The argument presented in both cases based on the inequality v > c or v < c, demonstrates that the relation in an ordered pairs of real numbers. Thus we cannot select these two numbers in the complex plane.

3. BEYOND THE SPEED LIMIT OF LIGHT AND RELATIVISTIC QUANTUM VIEW OF THE COMPLEX MATTER SPACE

Having a new unified presentation of mass and charge in the form of complex matter will bring a new question for us. How do we interpret the principles of Relativistic Quantum Mechanics?

Antimatter: Substance composed of elementary particles having the mass and electric charge of ordinary matter (such as electrons and protons) but for which the charge and related magnetic properties are opposite in sign. The existence of antimatter was predicted by the electron theory of P. A. M. Dirac. In 1932 the positron (antielectron) was detected in cosmic rays, followed by the antiproton and the antineutron detected through the use of particle accelerators. Positrons, antiprotons, and antineutrons, collectively called antiparticles, are the antiparticles of electrons, protons, and neutrons, respectively. When matter and antimatter are in close proximity, annihilation occurs within a fraction of a second, releasing large amounts of energy. The (relativistic) mass of an object measured by an observer in the xyz-frame is given by

$$M(t) = \frac{M_{\text{rest}}}{\sqrt{1-\left(\frac{v}{c}\right)^2}} = \frac{m_0 + iq_0}{\sqrt{1-\left(\frac{v}{c}\right)^2}} \tag{3.1}$$

By accepting the complex matter Postulate (I), the velocity in (3.1) can be any real or complex number. Mathematically v could be greater than the speed of light, but the mass m would become imaginary. Physically we would have to get to the speed of light first i.e. $v = c$, which gives us an undefined value for M. So we believe that nothing moves faster than the speed of light because we do not like observables to be imaginary.

A hypothetical particle that is traveling faster-than a light particle is consistent with the special theory of relativity. According to this theory, a free particle has an energy E and a momentum p which form a Lorentz four-vector. The length of this vector is a scalar, having the same value in all inertial reference frames. One writes

$$E^2 = c^2P^2 + m^2c^4 \tag{3.2}$$

Equation (3.2), where c is the speed of light and the parameter m^2 is a property of the particle, independent of its momentum and energy. Three cases may be considered: m^2 may be positive, zero, or negative. The case $m^2 > 0$ applies for atoms, nuclei, and the macroscopic objects of everyday experience. The positive root m is called the rest-mass. If $m^2 = 0$, the particle is called massless. A few of these are known: the electron neutrino, the muon neutrino, the photon, and the graviton. The name tachyons (after a Greek word for swift) has been given to particles with $m^2 < 0$.

Mayer in 2007, designed the energy momentum in a complex plane very well without acknowledging the idea of complex matter considering the speed of light $c = 1$ a natural unit (see [8]). The relation (3.2) can be interpreted by.

Energy in the complex plane: A vector form of the energy is demonstrated with the horizontal component OH = mc^2, and vertical component ZH = pc. The state of the energy and its magnitude can be described by

$$\boldsymbol{E} = \boldsymbol{m_0 c^2} + \boldsymbol{ipc} \tag{3.3}$$

$$\|E\| = \sqrt{\left(m_0 c^2\right)^2 + \left(pc\right)^2} \tag{3.4}$$

According to postulate (I), Figure 2 and Figure 3 demonstrate that the energy will have the two components of real and imaginary values. The energy for mass-less particles can be observed on the diagram when $\|E\| = pc$. When the real part of particle mass is not zero we will derive the energy, mass, and charge in relativistic mechanics. The norm (modulus) of the particle-like with speed faster than light can be described by

$$\|M\| = \frac{\|M_0\|}{\left\|\sqrt{1-\left(\frac{v}{c}\right)^2}\right\|} = \frac{\sqrt{m_0^2 + q_0^2}}{\sqrt{1-\left(\frac{v}{c}\right)^2}} = \frac{\sqrt{m_0^2 + q_0^2}}{i\sqrt{\left(\frac{v}{c}\right)^2 - 1}} = \gamma\sqrt{m_0^2 + q_0^2} = -i\bar{\gamma}\sqrt{m_0^2 + q_0^2}$$

where, the magnitude of the complex vector for particle and factor gamma is defined by:

$$\|M\| = \gamma\sqrt{m_0^2 + q_0^2} \quad \text{where } \gamma = \frac{1}{\sqrt{1-\left(\frac{v}{c}\right)^2}} \tag{3.5}$$

The following steps lead to the effects of complex matter theory on elementary quantum mechanics (see [9] and [10]).

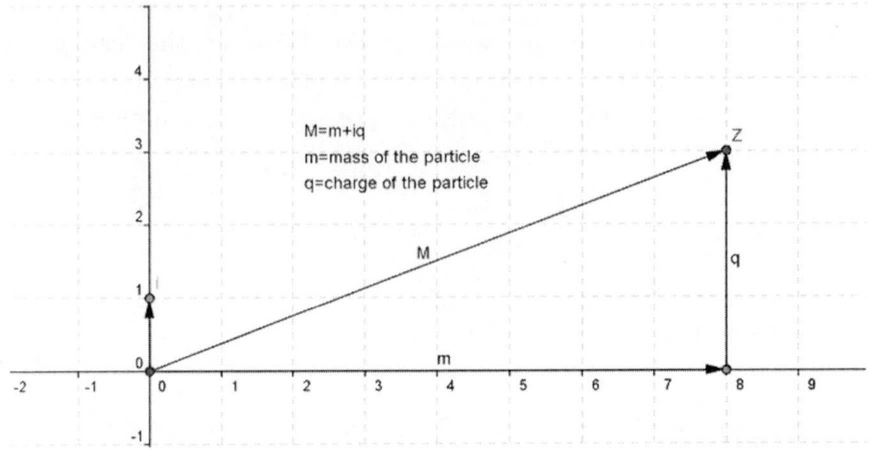

Figure 2. The state of the complex matter in a complex plane with two components (m, q) and it is denoted by $M = m + iq = re^{i\theta}$ where $r = \sqrt{m^2 + q^2}$, and $\theta = Arg\left(\frac{q}{m}\right)$.

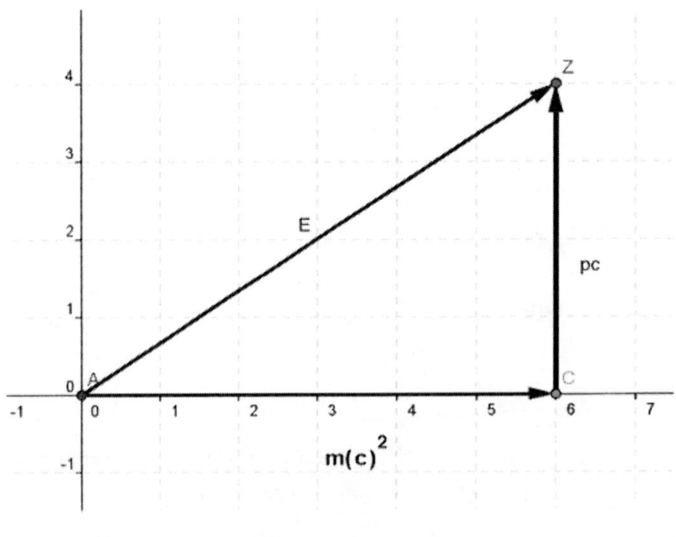

Figure 3. Energy in the complex plane: A vector form of the energy demonstrated with the horizontal component OH = mc^2, and vertical component ZH = pc. Energy vector $E = mc^2 + ipc$ with a magnitude $\|E\| = \sqrt{(m_0c^2)^2 + (pc)^2}$.

It is well known that the energy of a photon can be calculated if we assume that light is composed of particles.

The energy is proportional to the density of the photons particles, that is

1) Photon is a complex particle with zero mass. Energy of a package of photon particles is equal to energy of each particle multiply by the number of particles (frequency), that is

$$E = h \cdot f_q \tag{3.6}$$

Where, h is the Plank constant real number. The momentum can be calculated using.

2) Campton Effect when particle—like is a wave. Momentum of the of the photon wave will be equal to Energy/speed, that is

$$p_{real} = \frac{E}{c} = \frac{h \cdot f_q}{c} \tag{3.7}$$

By considering that velocity is equal to frequency multiplied by the wave length, and $c = \lambda \cdot f_q$ we will describe the momentum propagation of the photons as a wave by

$$p_{imag} = \frac{h \cdot f_q}{c} = \frac{h \cdot f_q}{\lambda \cdot f_q} = \frac{h}{\lambda} \tag{3.8}$$

3) Using de Broglie's Law we can find the energy of the wave (by 3.6) and replacing the frequency f_q is equal to the energy of the mass, that is $E_{imag} = h \cdot f_q = mc^2$.

Wave length $\lambda = \frac{h}{p} = \frac{h}{mv}$. Combining $\lambda = \frac{h}{p}$ and $E = h \cdot f = mc^2$ implies that: $w = \lambda f = \frac{h}{p} \cdot \frac{pc}{h} = c$. The wave frequency can be interpreted by $\frac{1}{\lambda} = f = \frac{p}{h}$.

This relation can also be extended to the Einstein mass- energy equivalence

$$\lambda = \frac{h}{p} = \frac{hc^2}{mvc^2} = \frac{hc^2}{E \cdot v} \Rightarrow \frac{p}{E} = \frac{v}{c^2} \text{ or } \frac{E}{p} = \frac{c^2}{v} \tag{3.9}$$

Consequently:

$$\frac{pc}{E} = \frac{v}{c} \tag{3.10}$$

Let E = mv²/2 and it is clear that $\mathrm{d}E/\mathrm{d}t = mv = p$. We use (7.4) to modify our definition of the angle alpha in (Figure 3): $\alpha = \sin^{-1}\left(\frac{pc}{E}\right) = \sin^{-1}\left(\frac{v}{c}\right)$.

4. MODIFYING THE EINSTEIN'S EQUATION IN COMPLEX MATTER SPACE

Using Postulate (I) let us describe a particle of a Complex Matter Space by a set of points $M(t) = m(t) + iq(t)$ where m is the mass and q is the charge at time t in the interval (a, b) for local real parameters a and b.

We will describe the instantaneous rates of changes of this complex matter by $\frac{\mathrm{d}M}{\mathrm{d}t} = \frac{\mathrm{d}m}{\mathrm{d}t} + i\frac{\mathrm{d}q}{\mathrm{d}t}$.

Complex Matter-Instantaneous Rate of Changes: Given a complex matter in the Equation (3.1) then the instantaneous rate of change satisfies the following relation

$$\frac{\mathrm{d}M}{\mathrm{d}t} = \frac{vM}{\left(c^2 - v^2\right)} \frac{\mathrm{d}v}{\mathrm{d}t} \text{ or } \mathrm{d}v = \frac{\left(c^2 - v^2\right)}{vM} \mathrm{d}M \tag{4.1}$$

Proof: Assuming $M = m + iq$ and taking the derivative of (3.1)

$\frac{\mathrm{d}M}{\mathrm{d}v}=\frac{vM_0}{c^2}\left(\frac{c^2}{c^2-v^2}\right)^{3/2}=\frac{vM_0}{c^2-v^2}\sqrt{\frac{c^2}{c^2-v^2}}$ implies

This will conclude the relation (4.1) and also $\mathrm{d}v=\frac{\left(c^2-v^2\right)}{vM}\mathrm{d}M$.

Energy Equation in Complex Matter Space: The instantaneous rate of change of Energy in Complex Matter Space can be described by

$$\mathrm{d}E=c^2\mathrm{d}m+ic^2\mathrm{d}q \tag{4.2}$$

Justification: Let us assume, in the following approach, that the definition of energy and force for the Complex Matter Space is staying the same. Thus;

$$\mathrm{d}E=F\mathrm{d}x=\frac{\mathrm{d}p}{\mathrm{d}t}\mathrm{d}x=\mathrm{d}p\frac{\mathrm{d}x}{\mathrm{d}t}=(\mathrm{d}p)v \tag{4.3}$$

where, the velocity $v=\frac{\mathrm{d}x}{\mathrm{d}t}$ is a real vector function, but the momentum $\boldsymbol{p}=M\cdot\boldsymbol{v}$ will be a complex vector function. The differentiation of momentum will be

$$\mathrm{d}p=M\mathrm{d}v+v\mathrm{d}M \tag{4.4}$$

Substitute (4.4) in the relation (4.3)

$$\mathrm{d}E=\left[M\mathrm{d}v+v\mathrm{d}M\right]v=v^2\mathrm{d}M+vM\frac{\left(c^2-v^2\right)\mathrm{d}M}{vM}$$

Simplifying the right hand side concludes

$$\mathrm{d}E=c^2\mathrm{d}M=c^2\mathrm{d}m+ic^2\mathrm{d}q \tag{4.5}$$

To explain the relativistic energy in Complex Matter Space, we use the trigonometric values in the complex plane (using the relation (3.10) and Figure 3);

$\alpha = \sin^{-1}\left(\frac{pc}{E}\right) = \sin^{-1}\left(\frac{v}{c}\right)$ and assume $\beta = \frac{v}{c}$

then $\cos(\alpha) = \sqrt{1-\sin^2(\alpha)} = \sqrt{1-\beta^2}$

When the real part of particle mass is not zero we will derive the energy, mass, and charge in relativistic mechanics.

The complex point E can be described in the following form

$$\boldsymbol{E} = \boldsymbol{mc}^2 + \boldsymbol{ipc} = \|E\| e^{i\alpha} = \sqrt{(m_0c^2)^2 + (pc)^2}\,(\cos(\alpha) + i\sin(\alpha))$$

Replacing cosine and sine of angle alpha from the previous step, we will get an energy state description

$$\boldsymbol{E} = \sqrt{(m_0c^2)^2 + (pc)^2}\left(\sqrt{1-\left(\frac{v}{c}\right)^2} + i\frac{v}{c}\right) \tag{4.6}$$

This is a correction of the conclusion presented in a paper by Mayer in 2007 (see [11]). The relation (4.6) is the state of the energy of a particle with the real rest mass moving faster than light.

5. STATE OF THE MOMENTUM IN COMPLEX MATTER SPACE

If the charge in wave-like particle is zero, then dividing the real part of energy by real part of momentum will produce the following relations

$$p_{real} = mv = \frac{m_0 v}{\sqrt{1-\left(\frac{v}{c}\right)^2}} \text{ and } E_{real} = mc^2 = \frac{m_0 c^2}{\sqrt{1-\left(\frac{v}{c}\right)^2}} \tag{5.1}$$

and we obtain $\frac{p_{real}}{E_{real}} = \frac{v}{c^2}$ which implies the relative velocities $\frac{v}{c} = \left\|\frac{pc}{E}\right\|$.

Using Figure 2 and the relation (3.10)

$$\frac{v}{c} = \frac{pc}{E} = \sin(\alpha) = \beta \tag{5.2}$$

The State of the Complex Matter: Using relations (5.2), one can find the value of gamma from (2.2)

$$\|M\| = \gamma\sqrt{m_0^2 + q_0^2} = \sqrt{m_0^2 + q_0^2}\sec(\alpha) \text{ where } \alpha = \sin^{-1}\left(\frac{v}{c}\right) \tag{5.3}$$

State of the Complex Energy: The complex point E can be described in the following state of energy in the complex model $\boldsymbol{E} = \boldsymbol{mc}^2 + \boldsymbol{ipc} = \|\boldsymbol{E}\|e^{i\alpha}$ Þ

$$\boldsymbol{E} = \sqrt{\left(m_0 c^2\right)^2 + (pc)^2}\left(\cos(\alpha) + i\sin(\alpha)\right) \tag{5.4}$$

Thus, from the relation (5.4) one can conclude the real and imaginary components of energy to be

$$E_{real} = \sqrt{\left(m_0 c^2\right)^2 + (pc)^2}\cos(\alpha), \text{ and } E_{imag} = \sqrt{\left(m_0 c^2\right)^2 + (pc)^2}\sin(\alpha) \tag{5.5}$$

(see [12]).

The State of the Momentum of Complex Matter: Assume that $M_0 = (m_0, q_0)$ represents the rest point at the initial time with two components. According to postulate (II), a moving particle in the complex plane at the

speed $v = \|V(t)\| = \left\|\frac{dx}{dt}\right\|$ on the one dimensional x-axis, will be a real function of time. The momentum at time t

can be described by

$$\|p\| = rv = \|M\|v = \frac{\|M_0\|v}{\sqrt{1-\frac{v^2}{c^2}}} = \frac{\sqrt{m_0^2+q_0^2}\,v}{\sqrt{1-\frac{v^2}{c^2}}} = v\sqrt{m_0^2+q_0^2}\sec(\alpha) \tag{5.6}$$

where, $\frac{1}{\sqrt{1-\left(\frac{v}{c}\right)^2}} = \frac{1}{\sqrt{1-\beta^2}} = \frac{1}{\cos(\alpha)} = \sec(\alpha)$. As a result the wave length can be described

$$\lambda = \frac{h}{\|p\|} = \frac{h}{v\sqrt{m_0^2+q_0^2}\sec(\alpha)} = \frac{h\cdot\cos(\alpha)}{v\sqrt{m_0^2+q_0^2}} \tag{5.7}$$

6. DISCUSSIONS AND CONCLUSION

According to the original postulate, we determine a point in a complex plane by $M = \|M\|e^{i\theta}$ where, $\theta = Arg\left(\frac{q}{m}\right)$. One can explore and study the implication of this assumption, and demonstrate the consistency with other phenomenon, theories, or experiments.

The existence of a particle traveling faster than light has been presented by many authors including, Sudarshan who discussed the Physics of complex mass (see [12]).

The idea of negative matter appears in past theories of matter that have now been abandoned (see [13]). Using the once popular vortex theory

10. John, W. and Wojceich, Z. (1983) The Quantum Postulate and the Recent Development of Atomic Theory. In: Quantum Theory and Measurement, Princeton University Press, Princeton, 111.

11. Reza, A. (2014) Quantum Complex Matter Space. International Journal of Theoretical and Mathematical Physics, 4, 159-163.

12. Zhang, T.X. (2008) Electric Charge as a form of Imaginary Energy. Progress in Physics, 2.

13. Sudarshan, E.C.G. (1968) A New Formulation of Relativistic Quantum Theory of Fields with Applications to Particles Traveling Faster Than Light. Proceedings of Nobel Symposium on Elementary Particle Theory.

14. Taylor, T.J.R., Zafirots, C.D. and Dubson, M.A. (2004) Modern Physics for Scientists and Engineers. 2nd Edition, Pearson-Printice Hall.

6

GENERALIZED PATH INTEGRAL TECHNIQUE: NANOPARTICLES INCIDENT ON A SLIT GRATING, MATTER WAVE INTERFERENCE

Valeriy I. Sbitnev

St.-Petersburg Nuclear Physics Institute, NRC Kurchatov Institute, Gatchina, Russia

1. INTRODUCTION

One of the crises of contemporary mathematics belongs in part to the subject of the infinite and infinitesimals [1]. It originates from the barest necessity to develop a rigorous language for description of observable physical phenomena. It was a time when foundations of integral and differential calculi were developed. A theoretical foundation for facilitation of understanding of classical mechanics is provided by the concepts of absolute time and space originally formulated by Sir Isaac Newton [2]. Space is distinct from body. And time passes uniformly without regard to whether anything happens in the world. For this reason Newton spoke of absolute space and absolute time as of a "container" for all possible objects and events. The space-time container is absolutely empty until prescribed metric and a reference frame are introduced. Infinitesimals are main tools of differential calculus [3, 4] within chosen reference frames.

Infinitesimal increment being a cornerstone of theoretical physics has one receptee default belief, that increment δV tending to zero contains a

of gravity, the possibility of matter with negative gravity was discussed by William Hicks in the 1880s.

In 1928, British physicist Paul A. M. Dirac generalized Einstein's famous equation $E = mc^2$ by considering negative mass. This hypothesis allowed for the existence of anti-particles in our universe.

These anti-particles are, literally, mirror images of normal or real matter. Each anti-particle has the same mass as its corresponding particle, but the electrical charges are reversed. Examples of antimatter discoveries of the 20th century are: 1) Positrons—Electrons with a positive instead of negative charge; 2) Anti-protons—Protons that have a negative instead of the usual positive charge; 3) Anti-atoms—Pairing together positrons and antiprotons (see [14]).

Scientists at the European Organization for Nuclear Research, created the first anti-atom. Nine anti-hydrogen atoms were created, each lasting only 40 nanoseconds. As of 1998, CERN researchers were pushing the production of anti-hydrogen atoms to 2000 per hour.

Both particles involved in the interaction process are completely annihilated and create a very huge explosion that occurs when antimatter and matter interact transferring the entire mass of both objects into energy.

It is part of the dream unification theory to build a matter-antimatter reaction engine?

7. FUTURE RESEARCH

The quantum states of the particles can be studied in the Complex Matter Space through interactions with other fundamental particles.

The Electromagnetic Interaction, Weak Interaction, and Strong Interaction forces can be considered in the Complex Matter Space. The standard model in quantum state does not consider Gravitation as part of the standard model.

Reza Ahangar The Complex Matter Space model can include all of the interactions as part of the model including the Weak Forces.

REFERENCES

1. Naber George, L. (1992) The Geometry of Minkowski Spacetime. An Introduction to the Mathematics of Special Theory of Relativity. Dover.
2. David, H. and Rensick (with the Assistance of Farrell Edwards and John Merrill) (1970) Fundamentals of Physics. John Wiley & Sons, Inc., Hoboken.
3. Dirac, P.A.M. (1928) The Quantum Theory of Electron. Proceedings of the Royal Society A, 117, 610-624.
4. Sudarshan, E.C.G. (1970) Physics of Complex Mass Particles. University of Texas, Austin.
5. Gerald, K. (2003) Quantum Physics, Relativity, and Complex Space Time: Toward a New Synthesis. (Originally published in 1990 by North-Holland, Amsterdam)
6. Richard, F. (1963) The Feynman Lectures on Physics. Vol. III, Addison Wesley, Reading, 1-6.
7. Mehran, R. (2010) Imaginary Relativity. Wseas Transaction on Communications, 9.
8. Mayer Alexander, F. (2007) Wave Energy in Quantum Mechanics. Journal of Physics, Conference Series, 70, Article ID: 012013. http://iopscience.iop.org/1712-6596/70/1/02013.
9. Wall Ernst, L. (1995) The Vortex Theory of the Electron and its de Broglie Waves Pulse. Institute for Basic Research, Palm Harbor. www.gwphysics.com www.tachyonmodel.com

lot of events to be under consideration. Probability of detection of a particle within this infinitesimal volume ρ(r⃗)δVρ(r→)δV is adopted as a smooth differentiable function with respect to its argument. From experience we know that for reproducing the probability one needs to accumulate enormous amount of events occurring within this volume. On the other hand we know, that as δV tends to zero we lose information about amount of the events. What is more, the information becomes uncertain. It means the infinitesimal increment being applied in physics faces with a conflict of depth of understanding physical processes on such minuscule scales. This trouble is avoided in quantum mechanics by proclamation that infinitesimal increments are operators, whereas observables are averaged on an ensemble.

In light of classical views Newton maintained the theory that light was made up of tiny particles, corpuscles. They spread through space in accordance with law of classical mechanics. Christian Huygens (a contemporary of Newton), believed that light was made up of waves vibrating up and down perpendicularly to direction of the light propagation. It comes into contradiction with Newtonian idea about corpuscular nature of light. Huygens was a proclaimer of wave mechanics as opposite to classical mechanics [5].

We abstain from allusion to physical vacuum but expand Huygensian idea to its logical completion. Let us imagine that all Newtonian absolute space is not empty but is populated everywhere densely by Huygensian vibrators. The vibrators are silent at absence of wave propagating through. But as soon as a wave front reaches some surface all vibrators on this surface begin to radiate at a frequency resonant with that of incident wave. From here it follows, that in each point of the space there are vibrators with different frequencies ranging from zero frequency up to infinite. All are silent in absence of an external wave perturbance. Thus, the infinitesimal volume δV is populated by infinite amount of the vibrators with frequencies ranging from zero to infinity. They are virtual vibrators facilitating propagation of waves through space.

Let us return to our days. One believes that besides matter and physical fields there is nothing more in the universe. Elementary particles are only a building material of "eternal and indestructible" substance of the cosmos. However we should avow that all observed matter and physical fields, are not the basis of the world, but they are only a small part of the total quantum reality. Physical vacuum in this picture constitutes a basic part of this reality. In particular, modern conception of the physical vacuum covers Huygens's idea perfectly. All space is fully populated by virtual particle-antiparticle pairs situated on a ground level. Such a particle-antiparticle pair has zero mass, zero charge, and zero magnetic moment. Famous Dirac' sea (Dirac postulated that the entire universe is entirely filled by particles with negative energy) is boundless space of electron-positron pairs populated everywhere densely - each quantum state possessing a positive energy is accompanied by a corresponding state with negative energy. Electron has positive mass and positron has the same mass but negative; electron has negative charge and positron has the same positive charge; when electron and positron dance in pair theirs magnetic moments have opposite orientations, so magnetic moment of the pair is zero.

Let electron and positron of a virtual pair rotate about mass center of this pair. Rotation of the pairs happens on a Bohr orbit. Energy level of the first Bohr orbit, for example, $mv^2/2$, is about 14 eV. Here m is electron mass and v is its velocity (on the first Bohr orbit the velocity is about 2.188 10^6m/s). Energy of the pair remains zero since positron has the same energy but with negative sign. Quantum fluctuations around this zero energy are as zero oscillations of electromagnetic field. Observe that, energy releasing of electron and positron from vacuum occurs at mc^2 = 1.022 MeV. So we see that there are about 7.5 10^4 Bohr orbits lying below this energy. That is, there is a vast scope for occupation of different Bohr orbits by the virtual electron-positron pairs.

A short outline given above is a basis for understanding of interference effects to be described below.

Ones suppose that random fluctuations of electron-positron pairs take place always. What is more, these fluctuations are induced by other pairs and by particles traveling through this random conglomerate. Edward Nelson has described mathematical models [6, 7] representing the above random fluctuations as Brownian motions of particles that are subjected by random impacts from particles populating aether

(Nelson's title of a lower environment). The model is viewed as a Markov process

$$dx(t) = b(x(t),t)dt + dw(t) \tag{1}$$

loaded by a Wiener term w with diffusion coefficient equal to $\hbar/2m$, where m is a mass of the particle and $\hbar$ is the reduced Planck constant. In this perspective Nelson has considered two Markov processes complementary to each other. One is described by forward-difference operator in the time, here$b(x(t),t)$ is a velocity calculated forward. And other equation is described by backward-difference operator in the time with the velocity $b^{\dagger}(x(t),t)$ calculated backward. In general $b(x(t),t) \neq b^{\dagger}(x(t),t)$ [6,7]. The two complementary processes, by means of transition to two new variables, real and imaginary, finally lead to emergence of the Schrödinger equation.

Nelson' vision that aether fluctuations look as Brownian movements of subparticles with $\hbar/2m$ being the diffusion coefficient of the movements, correlates with Feynman's ideas about quantum fluctuations of virtual particles in vacuum [8]. The Feynman path integral is akin to the Einstein-Smoluchowski integral equation [9]. The latter computes transition probability density. We shall deal with the modified Feynman path integral loaded by a temperature multiplied by the Boltzmann constant. At that, probability amplitude stays as a fundamental mathematical object at all stages of computations. As an example we shall consider emergence of interference patterns at scattering heavy particles on gratings. The particles are heavy in the sense that they adjoin to both realms, quantum and classical. They are nanoparticles. Such nanoparticles have masses about 100 amu and more, as, for example, fullerene molecules [10] shown in Fig. 1. It is remarkable that there are many experiments with such molecules showing interference patterns in the near field [11-18]. On the other hand these molecules are so large, that they behave themselves as classical particles at ordinary conditions.

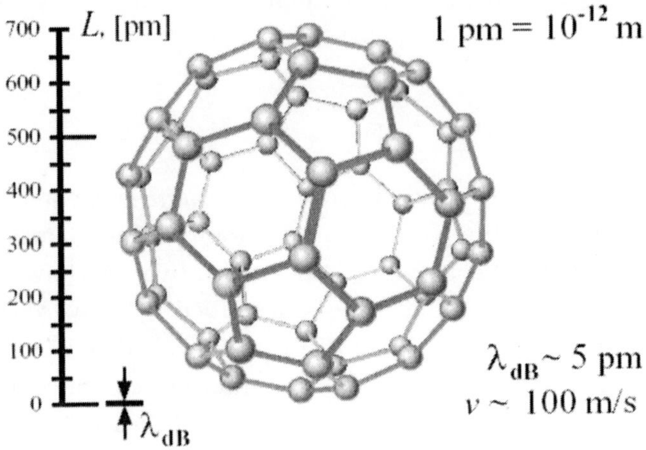

Figure 1. The fullerene molecule C_{60} consists of 60 carbon atoms. Its radius is about 700 pm. De Broglie wavelength of the molecule, λ_{dB}, is about 5 pm at a flight velocity v =100 m/s [18]. The molecules are prepared in a thermal emission gun which has temperature about 1000 K. It means that carbon atoms accomplish thermal fluctuations.

The article consists of five sections. Sec. 2 introduces a general conception of the path integral that describes transitions along paths both of classical and quantum particles. Here we fulfill expansions in the Taylor series of terms presented in the path integral. Depending on type of presented parameters we disclose either the Schrödinger equation or diffusion-drift equation containing extra term, osmotic diffusion. In the end of the section we compute passing nanoparticles through N-slit grating. Sec. 3 deals with interference patterns from the N-slit grating. Specifically, we study blurring of the Talbot carpet (an interference pattern emergent under special conditions imposed on the grating [19, 20]) arising under decoherence of incident on the grating nanoparticles. In Sec. 4 we find equations for computing the Bohmian trajectories. Also we compute variance of momenta along the trajectories. These computations lead to emergence of the uncertainty conditions. In concluding Sec. 5 we summarize results. For confirmation of existence of the Bohmian trajectories here we mention interference experiments with single silicon oil droplet [21].

2. GENERALIZED PATH INTEGRAL

Let many classical particles occupy a volume V and they move with different velocities in different directions. Let us imagine that there is a predominant orientation along which ensemble of the particles drifts. As a rule, one chooses a small volume δV in order to evaluate such a drift, Fig. 2. Learning of statistical mechanics begins with assumption that the volume should contain many particles.

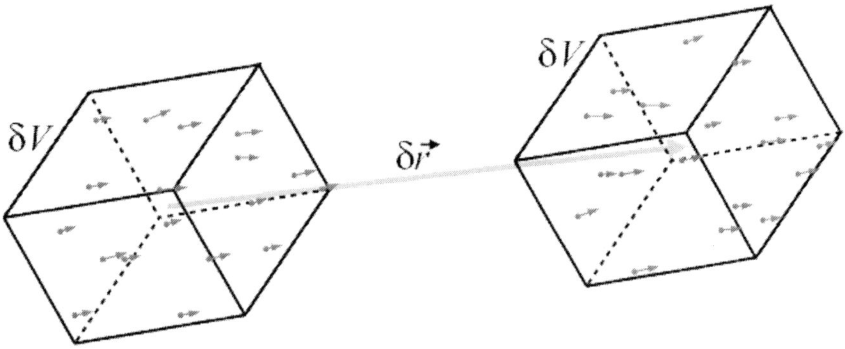

Figure 2. nfinitesimal volume δV contains many particles moving with different velocities having predominant orientation along blue arrow. The infinitesimal volume δV, as a mental image, is shifted along the same orientation.

The problem is to find transition probabilities that describe transition of the particle ensemble from one statistical state to another. These transient probabilities can be found through solution of the integral Einstein-Smoluchowski equation [9]. This equation in mathematical physics is known as Chapman-Kolmogorov [22-24] equation.. This equation looks as follows

$$p(\vec{r}_2,\vec{r}_1;t+\delta\tau) = \int_R p(\vec{r}_2,\vec{r};t+\delta\tau,t)\,p(\vec{r},\vec{r}_1;t)\,dV. \tag{2}$$

This equation describes a Markovian process without memory. That is, only previous state bears information for the next state. Integration here is fulfilled over all a working scene R, encompassing finite space volume. Infinitesimal volume δV in the integral tends to zero. It should be noted, however, that this volume should contain as many particles as possible for getting a satisfactory statistical pattern. One can see that this claim enters in conflict with the assumption $\delta V \to 0$.

Next we shall slightly modify approach to this problem. Essential difference from the classical probability theory is that instead of the probabilities we shall deal with probability amplitudes. The transition amplitudes can contain also imaginary terms. They bear information about phase shifts accumulated along paths. In that way, a transition from an initial state q⃗ 0q→0 to a final state q⃗ 1q→1 through all intermediate positions q⃗ xq→x given on a manifold R^3 (see Fig. 2) is represented by the following path integral

$$\psi(\vec{q}_1,\vec{q}_0;t+\delta t)=\int_R K(\vec{q}_1,\vec{q}_x;t+\delta t,t)\psi(\vec{q}_x,\vec{q}_0;t)D^3 q_x. \tag{3}$$

Here function $\psi(...)$ is a probability amplitude. Probability density $p(...)$, in turn, is represented by square of modulo of the probability amplitude, namely $p(...)=|\psi(...)|^2$. Integral kernel $K(\vec{q}_1,\vec{q}_x;t+\delta t,t)$ represents the transition amplitude from an intermediate state q⃗ xq→x to a final state q⃗ 1q→1. It is called propagator [8, 25, 26]. We suppose that the propagator has the following standard form

$$K(\vec{q}_1,\vec{q}_x;t+\delta t,t)=\frac{1}{A}\exp\left\{-\frac{L(\vec{q}_x,\dot{\vec{q}}_x)\delta t}{\Gamma}\right\}, \tag{4}$$

where denominator Γ under ex ponent is a complex-valued quantity, i.e., $\Gamma = \beta+\mathbf{i}\hbar$. The both parameters, β and the reduced Planck constant $\hbar$, have dimensionality of energy multiplied by time. From here it follows that $\beta = 2k_B T\delta t$. Here k_B is Boltzmann constant and T is temperature. So, we can write

$$\Gamma = 2k_B T\delta t + \mathbf{i}\hbar. \tag{5}$$

Factor 2 at the first term is conditioned by the fact that the kernel K relates to transitions of the probability amplitude ψ, not the probability *p*. Observe that a fullerene molecular beam for interference experiment prepared in a Knudsen cell at T=1070 K [27] spreads further in a vacuum. That is, fullerene molecules keep this temperature. From here it follows, that thermal fluctuations of carbon atoms from equilibrium occur at that temperature as the fullerene molecules propagate further within the vacuum chamber. One can see that at T=1000 K the term $2k_BT\delta t$ may be about $\hbar$ if δt is about 10^{-14} s.

Next let us imagine that particles pass through a path length one by one. That is, they do not collide with each other along the path length. The particles are complex objects, however. They are nanoparticles. Fullerene molecule, for example, contains 60 carbon atoms, Fig. 1. Conditionally we can think that the atoms are connected with each other by springs simulating elastic vibrations. In this view the Lagrangian L(q→x,q→˙x) can be written in the following form

$$L(\vec{q}_x,\dot{\vec{q}}_x)=\sum_{k=1}^{N}\left[\frac{m}{2}\frac{(\vec{q}_{1,k}-\vec{q}_{x,k}))^2}{\delta t^2}-U(\vec{q}_{x,k})\right]. \tag{6}$$

Here N is amount of atoms, constituent complex molecule, and m is mass of a single atom. By supposing that there are no quantum permutations between atoms we may expand the Lagrangian further

$$\begin{aligned}L(\vec{q}_x,\dot{\vec{q}}_x)&=\frac{m}{2}\sum_{k=1}^{N}\frac{(\vec{q}_{1,k}-\vec{q}_1+\vec{q}_1-\vec{q}_{x,k}+\vec{q}_x-\vec{q}_x)^2}{\delta t^2}-\sum_{k=1}^{N}U(\vec{q}_{x,k})\\&=\frac{m_N}{2}\frac{(\vec{q}_1-\vec{q}_x)^2}{\delta t^2}-m_N\frac{(\vec{q}_1-\vec{q}_x)}{\delta t}\frac{1}{N}\sum_{k=1}^{N}\frac{(\vec{\delta}_{1,k}-\vec{\delta}_{x,k})}{\delta t}+\frac{m}{2}\sum_{k=1}^{N}\frac{(\vec{\delta}_{1,k}-\vec{\delta}_{x,k})^2}{\delta t^2}-\sum_{k=1}^{N}U(\vec{q}_{x,k}).\end{aligned} \tag{7}$$

Here we admit that q⃗ xq→x and q⃗ 1q→1 are coordinates of center of mass of that complex molecule in intermediate and final positions. And *mN* = *Nm* is a mass of the molecule referred to its center of mass. Small deviations $\vec{\delta}_{1,k}=(\vec{q}_1-\vec{q}_{1,k})$ and $\vec{\delta}_{x,k}=(\vec{q}_x-\vec{q}_{x,k})$ are due to oscillations of the *k*th atoms with respect to the center of mass.

Let us consider the second row in Eq. (7). First of all we note that the term $(\vec{q}_1 - \vec{q}_x)/\delta t$ represents a velocity of movement of the center of mass. In such a case the first term is a kinetic energy of the center of mass. The second term represents product of the center mass velocity on an averaged velocity of partial oscillations of atoms constituting this molecule. The averaged velocity we believe vanishes because of conservation of total momentum. The third term represents a thermal kinetic energy of partial oscillating atoms constituting this molecular object. This energy is small enough. But it is sufficient to exhibit itself in the Casimir effect. The last term is a total potential energy $U(\vec{q}_x)$ in the point $\vec{q}_x$.

The path integral (3) contains functions depending only on coordinates $\vec{q}_0$, $\vec{q}_1$, $\vec{q}_x$ relating to positions of the center of mass. Whereas the Lagrangian (6) gives description for behavior of each atom constituent the complex molecule. Here we shall suppose that oscillations of all atoms are noncoherent. And consequently they do not give contribution to interference effect on output. We believe that these oscillations provide a thermal noise. And next we shall replace this oscillating background by a corresponding thermal term. For this reason, we believe that along with the reduced Planck constant $\hbar$ the parameter $2k_B T\delta t$ in Eq. (5) can be different from zero as well.

2.1. Expansion Of The Path Integral

The next step is to expand terms, ingoing into the integral (3), into Taylor series. The wave function written on the left is expanded up to the first term

$$\psi(\vec{q}_1, \vec{q}_0; t + \delta t) \approx \psi(\vec{q}_1, \vec{q}_0; t) + \frac{\partial \psi}{\partial t}\delta t. \tag{8}$$

As for the terms under the integral, here we preliminarily make some transformations. We define a small increment

$$\vec{\xi} = \vec{q}_1 - \vec{q}_x \Rightarrow D^3 q_x = -D^3 \xi. \tag{9}$$

The Lagrangian (7) is rewritten, in such a case, in the following view

$$L(\vec{q}_x, \dot{\vec{q}}_x) = \frac{m_N}{2} \frac{\xi^2}{\delta t^2} - m_N \frac{\xi}{\delta t} \frac{1}{N} \sum_{k=1}^{N} \frac{(\vec{\delta}_{1,k} - \vec{\delta}_{x,k})}{\delta t} + \frac{m}{2} \sum_{k=1}^{N} \frac{(\vec{\delta}_{1,k} - \vec{\delta}_{x,k})^2}{\delta t^2} - U(\vec{q}_x) \tag{10}$$

Here $U(\vec{q}_x)$ is sum of all potentials $U(\vec{q}_{x,k})$ given in the center of mass. Further we shall deal with the path integral (3) where the kernel *K* contains the Lagrangian given from Eq. (10). The under integral function

$\psi(\vec{q}_x, \vec{q}_0; t) = \psi(\vec{q}_1 - \vec{\xi}, \vec{q}_0; t)$ is subjected to expansion into the Taylor series up to the second terms of the expansion

$$\psi(\vec{q}_1 - \vec{\xi}, \vec{q}_0; t) \approx \psi(\vec{q}_1, \vec{q}_0; t) - \left(\nabla \psi, \vec{\xi}\right) + \nabla^2 \psi \cdot \xi^2 / 2 \tag{11}$$

The potential energy $U(\vec{q}_x) = U(\vec{q}_1 - \vec{\xi})$ is subjected to expansion into the Taylor series by the small parameter ξ⃗ ξ→also. Here we restrict themselves by the first two terms of the expansion, $U(\vec{q}_1) - (\nabla U(\vec{q}_1), \vec{\xi})$.

Taking into account the expressions (8)-(11) and substituting theirs into Eq. (3) we get

$$\psi(\vec{q}_1, \vec{q}_0; t) + \frac{\partial \psi}{\partial t} \delta t = -\frac{1}{A} \int_{R^3} \exp\left\{ -\frac{1}{\Gamma} \left(\frac{m_N}{2} \frac{\xi^2}{\delta t} - \xi \underbrace{\frac{m_N}{N} \sum_{k=1}^{N} \frac{(\vec{\delta}_{1,k} - \vec{\delta}_{x,k})}{\delta t}}_{(a)} + \underbrace{\left[\frac{m}{2} \sum_{k=1}^{N} \frac{(\vec{\delta}_{1,k} - \vec{\delta}_{x,k})^2}{\delta t^2} \right]}_{(b)} \delta t \right.\right.$$
$$\left.\left. - \left(U(\dot{q}_1) - (\nabla U(\dot{q}_1), \vec{\xi})\right) \delta t \right) \right\} \left(\psi(\dot{q}_1, \dot{q}_0; t) - \left(\nabla \psi, \vec{\xi}\right) + \nabla^2 \psi \cdot \xi^2 / 2 \right) D^3 \xi. \tag{12}$$

First, we consider terms enveloped by braces (a) and (b): (a) here displacement $(\vec{\delta}_{1,k}-\vec{\delta}_{x,k})$ divided by δt is a velocity $\vec{v}_k$ of k*th* atom at its deviation from a steady position relative to the center of mass. Summation through all deviations of atoms divided by *N* gives averaged velocity, $\vec{v}$, of all atoms with respect to the center of mass. This averaged velocity, as we mentioned above, vanishes. The velocity can be nonzero only in a case when external forces push coherently all atoms. This case we do not consider here. (b) this term is a thermal kinetic energy, *TN*, of the atoms oscillating about the center of mass. Observe that energy of thermal fluctuations, *TN*, is proportional to $k_B T$. Because of its presence in the propagator intensity of an interference pattern diminishes in general. Further we shall add this term into the potential energy as some constant component.

In the light of the above observation we rewrite Eq. (12) as follows

$$\psi(\vec{q}_1,\vec{q}_0;t)+\frac{\partial\psi}{\partial t}\delta t=-\frac{1}{A}\int_{R^3}\exp\left\{-\frac{m_N}{2\Gamma}\frac{\xi^2}{\delta t}\right\}\left(1+\underbrace{m_N\vec{v}}_{(a)}\frac{\vec{\xi}}{\Gamma}+\underbrace{T_N}_{(b)}\frac{\delta t}{\Gamma}+\left(U(\vec{q}_1)-(\nabla U(\vec{q}_1),\vec{\xi})\right)\frac{\delta t}{\Gamma}\right)$$
$$\times\left(\psi(\vec{q}_1,\vec{q}_0;t)-(\nabla\psi,\vec{\xi})+\nabla^2\psi\cdot\xi^2/2\right)D^3\xi.$$

(13)

Here we have expanded preliminarily exponents to the Taylor series up to the first term of the expansion. Exception relates to the term exp{ -*mN* $\xi^2/2\Gamma\delta t$} which remains in its original form. This exponent integrated over all space R^3 results in

$$-\frac{1}{A}\int_{R^3}\exp\left\{-\frac{m_N}{2\Gamma}\frac{\xi^2}{\delta t}\right\}D^3\xi=-\frac{1}{A}\left(\frac{2\pi\Gamma\delta t}{m_N}\right)^{3/2}=1\quad\Rightarrow\quad A=-\left(\frac{2\pi\Gamma\delta t}{m_N}\right)^{3/2}.$$

(14)

To derive outcomes of integration of terms containing factors $(\nabla\psi,\vec{\xi})$ and $\nabla^2\psi\cdot\xi^2/2$ we mention the following integrals [8]

$$\frac{1}{A}\int_{R^3}\exp\left\{-\frac{m_N}{2\Gamma}\frac{\xi^2}{\delta t}\right\}\xi D^3\xi = 0 \qquad \text{and} \qquad \frac{1}{A}\int_{R^3}\exp\left\{-\frac{m_N}{2\Gamma}\frac{\xi^2}{\delta t}\right\}\xi^2 D^3\xi = \frac{\Gamma}{m_N}\delta t$$

(15)

In the light of this observation let us now solve integral (13) accurate to terms containing δt not higher the first order:

$$\psi(\vec{q}_1,\vec{q}_0;t)+\frac{\partial\psi}{\partial t}\delta t = \psi(\vec{q}_1,\vec{q}_0;t)+\frac{\Gamma}{2m_N}\nabla^2\psi\,\delta t+\frac{1}{\Gamma}(U(\vec{q}_1)-T_N)\psi\,\delta t.$$

(16)

We note that the term (v⃗ ,∇ψ) here is absent since we consider v⃗ =0v→=0, as was mentioned above. By reducing from the both sides the function $\psi(\vec{q}_1,\vec{q}_0;t)$ we come to the following differential equation

$$\frac{\partial\psi(\vec{q}_1,\vec{q}_0;t)}{\partial t} = \frac{\Gamma}{2m_N}\nabla^2\psi(\vec{q}_1,\vec{q}_0;t)+\frac{1}{\Gamma}(U(\vec{q}_1)-T_N)\psi(\vec{q}_1,\vec{q}_0;t). \tag{17}$$

The parameter

$$\frac{\Gamma}{2m_N} = \frac{2k_B T\delta t}{2m_N}+\mathbf{i}\frac{\hbar}{2m_N} \tag{18}$$

is seen to be as a complex-valued diffusion coefficient consisting of real and imaginary parts.

2.1.1. Temperature Is Zero

Let $k_B T\delta t = 0$. It means that $\Gamma = \mathrm{i}\hbar$. Also $TN = 0$. One can see that Eq. (17) is reduced to

$$\mathbf{i}\hbar\frac{\partial\psi(\vec{q}_1,\vec{q}_0;t)}{\partial t} = -\frac{\hbar^2}{2m_N}\Delta\psi(\vec{q}_1,\vec{q}_0;t)+U(\vec{q}_1)\psi(\vec{q}_1,\vec{q}_0;t). \tag{19}$$

It is the Schrödinger equation.

2.1.2. Temperature Is Not Zero

Let $k_B T\delta t >> \hbar$. We can suppose that $\Gamma = k_B T\delta t$. Eq. (17) takes a form

$$\frac{\partial\psi(\vec{q}_1,\vec{q}_0;t)}{\partial t} = D\Delta\psi(\vec{q}_1,\vec{q}_0;t) + \frac{\left(U(\vec{q}_1) - T_N\right)}{2m_N D}\psi(\vec{q}_1,\vec{q}_0;t). \tag{20}$$

Here

$$D = \frac{k_B T\delta t}{m_N} \tag{21}$$

is the diffusion coefficient. The coefficient has dimensionality of [length^2 time^{-1}]. It is a factor of proportionality representing amount of substance diffusing across a unit area in a unit time - concentration gradient in unit time.

We can see that Eq. (20) deals with the amplitude function ψ, not a concentration. However, a measurable function is $\rho=|\psi|^2$ - concentration having dimensionality of [(amount of substance) length^{-3}]. Let us multiply Eq. (20) from the left by 2ψ. First we note that the combination $2\psi\Delta\psi = 2\psi\nabla\psi^{-1}\psi\nabla\psi = 2(\psi\nabla\psi^{-1})(\psi\nabla\psi) + 2\nabla(\psi\nabla\psi)$ results in $-(1/2)(\nabla\ln(\rho)\cdot\nabla\rho) + \Delta\rho.$. As a result we come to a diffusion-drift equation describing diffusion in a space loaded by a potential field $(U(\vec{q}_1) - T_N)$:

$$\frac{\partial\rho}{\partial t} + \frac{D}{2}\left(\nabla\ln(\rho),\nabla\rho\right) = D\Delta\rho + \frac{(U(\vec{q}_1) - T_N)}{2m_N D}\rho \tag{22}$$

Extra term (D/2)∇ln(ρ) in this diffusion-drift equation is a velocity of outflow of the particles from volume populated by much more number of particles than in adjacent volume. The term −ln(ρ)−ln(ρ) is entropy of a particle ensemble. From here it follows that ∇ln(ρ)∇ln(ρ) describes inflow of the particles to a region where the entropy is small. Observe that the velocity

$$\vec{u} = D\frac{\nabla\rho}{\rho} = D\nabla\ln(\rho) \tag{23}$$

is an osmotic velocity required of the particle to counteract osmotic effects [6]. Namely, imagine a suspension of many Brownian particles within a physical volume acted on by an external, virtual in general, force. This force is balanced by an osmotic pressure force of the suspension [6]:

$$\vec{K} = k_B T\frac{\nabla\rho}{\rho}. \tag{24}$$

From here it is seen, that the osmotic pressure force arises always when density difference exists and especially when the density tends to zero. And vice-versa, the force disappears in extra-dense media with spatially homogeneous distribution of particles. As states the second law of thermodynamics spontaneous processes happen with increasing entropy. The osmosis evolves spontaneously because it leads to increase of disorder, i.e., with increase of entropy. When the entropy gradient becomes zero the system achieves equilibrium, osmotic forces vanish.

Due to appearance of the term $(D/2)\nabla\ln(\rho)$ in Eq. (22) the diffusion equation becomes nonlinear. It is interesting to note, running ahead, that this osmotic term reveals many common with the quantum potential, which is show further. Observe that the both expressions contain the term $\nabla\rho/\rho=\nabla\ln(\rho)\nabla\rho/\rho=\nabla\ln(\rho)$relating to gradient of the quantum entropy $SQ=-\ln(\rho)/2$ [28].

Reduction to PDEs, Eq. (19) and Eq. (22), was done with aim to show that the both quantum and classical realms adjoin with each other much more closely, than it could seem with the first glance. Further we shall return to the integral path paradigm [8] and calculate patterns arising after passing particles through gratings. We shall combine quantum and classical realms by introducing the complex-valued parameter $\Gamma = 2k_B T\delta t + i\hbar$.

2.2. Paths Through *N*-Slit Grating

Computation of a passing particle through a grating is based on the path integral technique [8]. We begin with writing the path integral that describes passing the particle through a slit made in an opaque screen that is situated perpendicularly to axis *z*, Fig. 3. For this reason we need to describe a movement of the particle from a source to the screen and its possible deflection at passing through the slit, seeFig. 4. At that we need to evaluate all possible deflections.

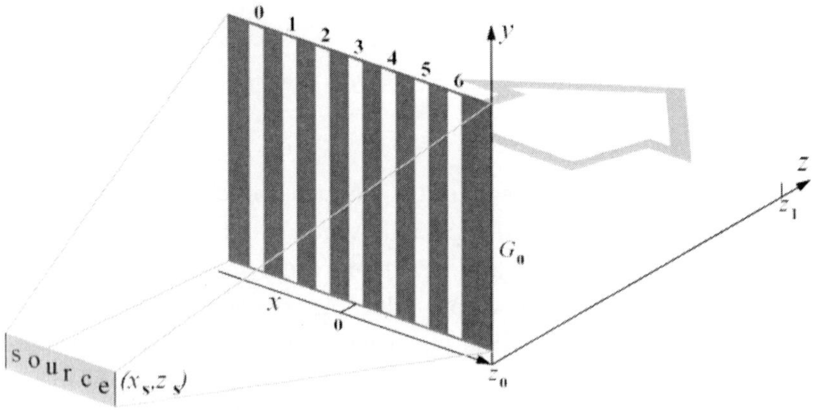

Figure 3. Interferometry from one grating G_0 situated transversely to a particle beam emitted from a distributed source.

We believe, that before the screen and after it, the particle (fullerene molecule, for example) moves as a free particle. Its Lagrangian, rewritten from Eq. (17), describes its deflection from a straight path in the following form

$$L = m_N \frac{\dot{x}^2}{2} - m_N(\dot{x}, \vec{v}) + T_N. \qquad (25)$$

The first term relates to movement of the center of mass of the molecule. So that *mN* is mass of the molecule and x˙x˙ is its transversal velocity, i.e., the velocity lies in transversal direction to the axis *z*. The second term is conditioned by collective fluctuations of atoms constituent this molecule. This term is nonzero when atoms have predominant fluctuations along axis *x*. For the sake of simplicity we admit that $\vec{v}$ is constant. The third term is a constant and comes from Eq. (17). It can be introduced into the normalization factor. For that reason we shall ignore this term in the following computations. A longitudinal momentum, p_z, is much greater than its transverse component [16, 17, 29] and we believe it is constant also. By translating a particle's position on a small distance $\delta x = (x_b - x_a) << 1$ for a small time $\delta t = (t_b - t_a) << 1$ we find that a weight factor of such a translation has the following form

$$\exp\left\{-\frac{L\delta t}{\Gamma}\right\} = \exp\left\{-\frac{m_N}{2\Gamma}\frac{(x_b - x_a)^2}{(t_b - t_a)} + \frac{m_N v}{\Gamma}(x_b - x_a)\right\}. \tag{26}$$

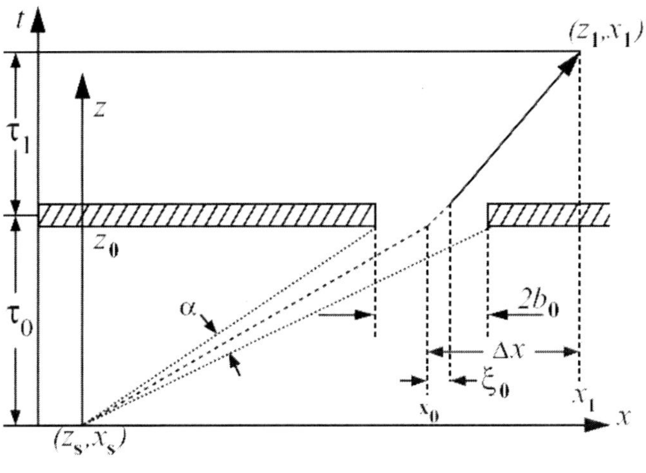

Figure 4. Passage of a particle along path $(z_s, x_s) \rightarrow (z_0, x_0) \rightarrow (z_1, x_1)$ through a screen containing one slit with a width equal to $2b_0$. Divergence angle of particles incident on the slit, α, tends to zero as the source is removed to infinity.

The particles flying to the grating slit along a ray α, Fig. 4, pass through the slit within a range fromx_0-b_0 to x_0+b_0 with high probability. The path integral in that case reads

$$\psi(x_1,x_0,x_s)=\int_{-b_0}^{b_0}K(x_1,\tau_0+\tau_1;x_0+\xi_0,\tau_0)K(x_0+\xi_0,\tau_0;x_s,0)d\xi_0. \qquad (27)$$

Integral kernel (propagator) for the particle freely flying is as follows [8, 26]

$$K(x_b,t_b;x_a,t_a)=\left[\frac{m_N(1-2v/v_{(a,b)})}{2\pi\Gamma(t_b-t_a)}\right]^{1/2}\exp\left\{-\frac{m_N(x_b-x_a)^2}{2\Gamma(t_b-t_a)}\left(1-2\frac{v}{v_{(a,b)}}\right)\right\} \qquad (28)$$

Here $v_{(a,b)}=(x_b-x_a)/(t_b-t_b)$ is a velocity of the molecule on a segment from *xa* to *xb*. And *v* is an average velocity of collective deflection of atoms constituent the molecule. It was defined in Eq. (12). We shall believe that the ratio *v/v (a,b)* is small enough. We can define a new renormalized mass $\{m\}_N=m_N(1-v/v_{(a,b)})$ and further we shall deal with this mass.

2.2.1. Passing Of A Particle Through Slit

By substituting the kernel (28) into the integral (27) we obtain the following detailed form

$$\psi(x_1,x_0,x_s)=\int_{-b_0}^{b_0}\left(\frac{\{m\}_N}{2\pi\Gamma\tau_1}\right)^{1/2}\exp\left\{-\frac{\{m\}_N(x_1-(x_0+\xi_0))^2}{2\Gamma\tau_1}\right\}\left(\frac{\{m\}_N}{2\pi\Gamma\tau_0}\right)^{1/2}\exp\left\{-\frac{\{m\}_N((x_0+\xi_0)-x_s)^2}{2\Gamma\tau_0}\right\}d\xi_0. \qquad (29)$$

The integral is computed within a finite interval [-b_0,+b_0]. Observe, that the integrating can be broadened from -∞ to +∞. But in this case we need to load the integral by the step function equal to unit within the finite interval [-b_0,+b_0] and it vanishes outside of the interval. The step function, that simulate a single slit, can be approximated by the following a set of the Gaussian functions [30]

$$G(\xi,b,\eta,K)=\frac{1}{\eta}\sqrt{\frac{2}{\pi}}\sum_{k=1}^{K}\exp\left\{-\frac{(K\xi-b(K-(2k-1)))^2}{2(b\eta)^2}\right\}. \tag{30}$$

Here parameter b is a half-width of the slit, real $\eta > 0$ is a tuning parameter, and K takes integer values. At $K = 1$ this form-factor degenerates to a single Gaussian function. And at $K \to \infty$ this function tends to an infinite collection of the Kronecker deltas which fill everywhere densely the step function. We rewrite Eq. (29) with inserting this form-factor

$$\psi(x_1,x_0,x_s)=\int_{-\infty}^{\infty} G(\xi_0,b_0)\frac{\{m\}_N}{2\pi\Gamma\sqrt{\tau_1\tau_0}}\exp\left\{-\frac{\{m\}_N}{2\Gamma}\left(\frac{(x_1-(x_0+\xi_0))^2}{\tau_1}+\frac{((x_0+\xi_0)-x_s)^2}{\tau_0}\right)\right\}d\xi_0. \tag{31}$$

We do not write parameters η and K in the Gaussian form-factor and for the sake of simplicity further we shall consider they equal to 1. That is, for simulating the slit we select a single Gaussian function.

2.2.2. Definition Of New Working Parameters

First we replace the flight times τ_0 and τ_1 by flight distances (z_0-z_s) and (z_1-z_0), see Fig. 4. This replacement reads

$$\begin{cases}\tau_0=(z_0-z_s)/v_z\\ \tau_1=(z_1-z_0)/v_z\end{cases} \tag{32}$$

where vz is a particle velocity along the axis z.

There is, however, one more parameter of time that is represented in definition of the coefficient $\Gamma = 2k_BT\delta t + i\hbar$. It is a small time increment δt. The parameter δt first appears in the path integral (3) as the time increment along a path. In accordance with the uncertainty principle δt should be more or equal to the ratio of $\hbar$, Planck constant, to energy of occurring events. In a case of a flying particle through vacuum it can be minimal energy of vacuum fluctuations (it is about energy of the first Bohr orbit of electron-positron pair that is about 14 eV). Evaluation gives $\delta t \sim 2.8\ 10^{-16}$ s. From here it follows, that $2k_BT\delta t$ is less than $\hbar$ on about one order at T = 1000 K (almost temperature of fullerene evaporation from the Knudsen cell [27]).

Emergence of the term $2k_BT\delta t$ can be induced by existence of quantum drag [31] owing to different conditions for quantum fluctuations both inside of the fullerene molecule and outside what can induce weak Casimir forces. Because of the weak Casimir force the quantum drag does not lead to decoherence at least in the near zone. However further we shall see that a weak washing out of the Talbot interference pattern is due to existence of this term.

Let us divide the parameter Γ by $\{m\}_N v_z$

$$\frac{\Gamma}{\{m\}_N v_z} = \frac{2k_BT}{\{m\}_N v_z}\delta t + \mathbf{i}\frac{\hbar}{\{m\}_N v_z}. \tag{33}$$

Here $p_z = \{m\}_N v_z$ is a particle momentum along axis z. We can define the de Broglie wavelength $\lambda_{dB} = h/p_z$ where $h = 2\pi\hbar$ is the Planck constant. Let us also define a length $\delta_T = 4\pi k_BT\delta t/(\{m\}_N v_z)$. In this view we can rewrite Eq. (33) as follows

$$\frac{\Gamma}{\{m\}_N v_z} = \frac{\delta_T}{2\pi} + \mathbf{i}\frac{\lambda_{dB}}{2\pi} = \frac{1}{2\pi}\Lambda. \tag{34}$$

The length δ_T tends to zero as $T \to 0$. At T = 1000 K and at adopted δt = 2.8 10^{-16} s we have $\delta_T \approx$ 0.4 pm. On the other hand, the de Broglie wavelength, λ_{dB}, evaluated for the fullerene molecule moving with the velocity v_z =100 m/s is about 5 pm [18]. So, we can see that the length δ_T is less of the de Broglie wavelength on about one order and smaller.

A signification of the length δT is that it determines decoherence of a particle beam. Decoherence of flying particles occurs the quickly, the larger δT. Observe that the length δT has a close relation with the coherence width - a main parameter in the generalized Gaussian Schell-model [32, 33].

3. WAVE FUNCTION BEHIND THE GRATING

Wave function from one slit after integration over ξ_0 from -∞ to +∞ has the following view [28]

$$\psi(x,z,x_0,x_s) = \sqrt{\frac{2}{\pi\Lambda\Sigma_0(z_0-z_s)}}\exp\left\{-\frac{\pi}{\Lambda}\left[\frac{(x-x_0)^2}{(z-z_0)}\left(1-\frac{\Xi_0^2}{\Sigma_0}\right)+\frac{(x_0-x_s)^2}{(z_0-z_s)}\right]\right\}. \tag{35}$$

Here argument of ψ-function contains apart x also z in order to emphasize that we carry out observation in the point (z,x), see Fig. 4. The factor $(2/\pi)^{1/2}$ comes from (30). Parameters Ξ_0 and $\Sigma_0$read

$$\Xi_0 = 1 - \frac{(x_0-x_s)(z-z_0)}{(z_0-z_s)(x-x_0)} \quad \text{and} \quad \Sigma_0 = \frac{(z-z_s)}{(z_0-z_s)} + \frac{\Lambda(z-z_0)}{2\pi b_0^2} \tag{36}$$

In order to simplify records here we omit subscript 1 at x and z - an observation point that is situated after the slit.

Let us consider that an opaque screen contains N_0 slits spaced through equal distance, d, from each other. Numeration of the slits is given as it is shown in Fig. 3, n_0 = 0,1,2, ..., N_0 – 1. Sum of all wave functions (35), each of which calculates outcome from an individual slit, gives a total effect in the point (z,x) where a detector is placed:

$$|\Psi_0(x,z,d,x_s,\Lambda)\rangle = \sum_{n_0=0}^{N_0-1}\psi\left(x,z,\left(n_0-\frac{N_0-1}{2}\right)d,x_x\right). \tag{37}$$

Probability density in the vicinity of the observation point (x,z) reads

$$p(x,z) = \langle \Psi_0(x,z,d,x_s,\Lambda) | \Psi_0(x,z,d,x_s,\Lambda) \rangle. \tag{38}$$

Calculation of the wave function (37) is fulfilled for the grating containing N_0 = 32 slits. Distance between slits is $d = 10^5 \lambda_{dB}$. So at λ_{dB} = 5 pm the distance is equal to 500 nm. Requirement $\lambda_{dB} << d$and N_0 tending to infinity together with a condition that the particle beam is paraxial, that is, x_s = 0 and $z_s \to -\infty$, provides emergence in the near-field of an interference pattern, named Talbot carpet [19,20]. Here a spacing along interference patterns is measured in the Talbot length

$$z_T = 2\frac{d^2}{\lambda_{dB}}, \tag{39}$$

which is a convenient natural length at representation of interference patterns. Since we restrict themselves by finite N_0 we have a defective carpet, which progressively collapses as a spacing from the slit increases. Fig. 5 shows the Talbot carpet, being perfect in the vicinity of the grating slit; it is destroyed progressively with increasing z_T. As for the Talbot carpet we have the following observation. We see that in a cross-section $z = z_T/2$ image reproduces radiation of the slits but phase-shifted by half period between them. At $z = z_T$ radiation of the slits is reproduced again on the same positions where the slits are placed. And so forth.

Evaluation of sizes of the interference pattern is given by ratio of the Talbot length to a length of the slit grating. In our case the Talbot length is z_T = 0.1 m. And length of the slit grating is about $N_0 d$ = 1.6 10^{-5} m. From here we find that the ratio is 6250. It means that the interference pattern shown inFig. 5 represents itself a very narrow strip.

Zigzag curves, drawn in the upper part of Fig. 5 by dark blue color, show Bohmian trajectories that start from the slit No. 15. One can see that particles prefer to move between nodes having positive interference effect and avoid empty lacunas. However the above we noted, that the ratio of the Talbot length to the length of the grating is about 6250 >> 1.

It means that really the Bohmian trajectories look almost as straight lines slightly divergent apart. Zigzag-like behavior of the trajectories is almost invisible. Such an almost feebly marked zigzag-like behavior may be induced by fluctuations of virtual particles escorting the real particle.

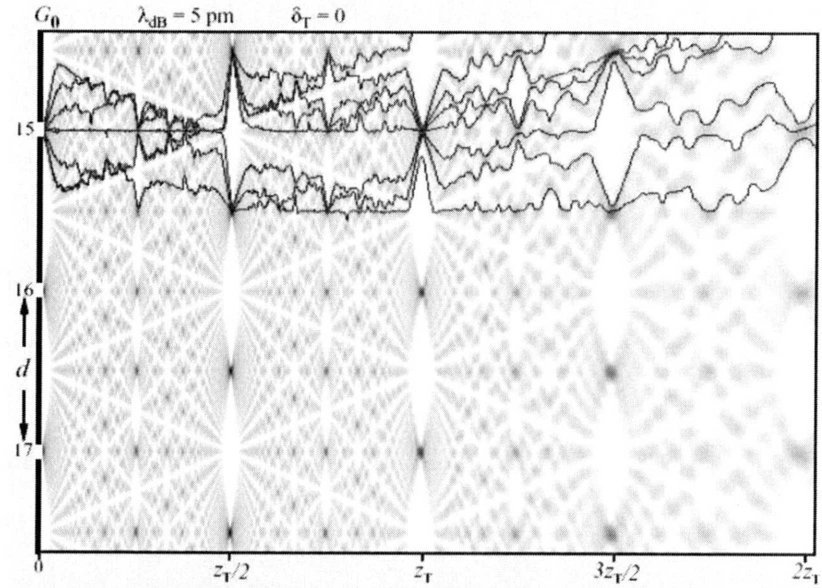

Figure 5. Interference pattern in the near field. It is shown only the central part of the grating containing N_0 = 32 slits, λ_{dB} = 5 pm, d = 500 nm, and δ_T = 0. In the upper part of the figure a set of the Bohmian trajectories, looking like on zigzag curves, drawn in dark blue color is shown.

As soon as we add the term $k_B T\delta t$ different from zero (T > 0 K) we observe blurring the interference pattern. The blurring is the stronger, the larger $k_B T\delta t$. For comparison see Figs. 6 and 7. Here instead of $k_B T\delta t$ we write a more convenient parameter, the coherence length δ_T. This length characterizes a dispersed divergence from initially tuned the de Broglie wavelength. Such a disperse medium can be due to quantum drag on the vacuum fluctuations. Here the Bohmian trajectories are not shown, since because of the thermal term $k_B T \delta t > 0$ a Brownian like scattering of the trajectories arises. This scattering we shall discuss later on.

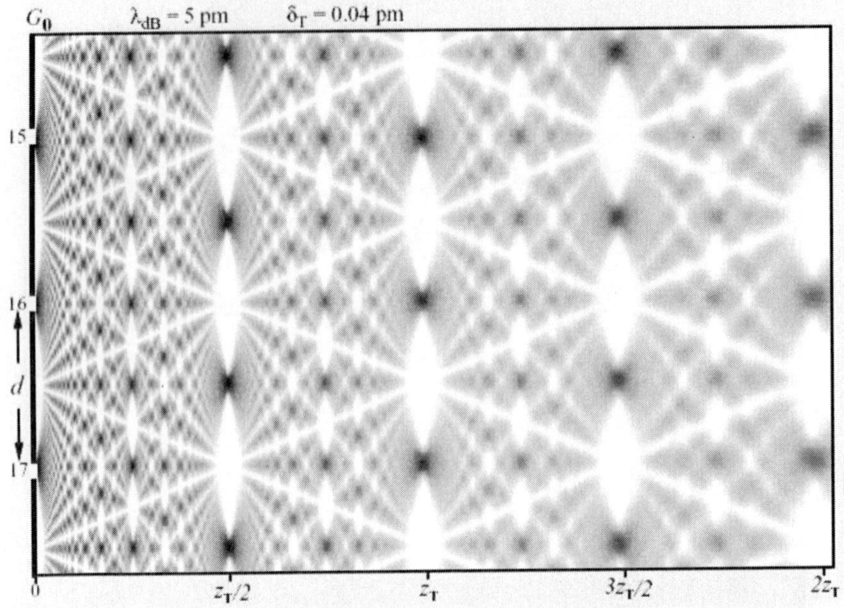

Figure 6. Blurred interference pattern in the near field. It is shown only the central part of the grating containing N_0 = 32 slits; d = 500 nm, δ_T = 0.04 pm << λ_{dB} = 5 pm.

We may think that the technical vacuum can be not perfect. It causes additional scattering of particles on residual gases. Because of this additional scattering the interference pattern can be destroyed entirely, as shown in Fig. 8.

Now let us draw dependence of the probability density $p(x, z)$ as a function of x at fixed z. In other words, we calculate interference fringes in a cross-section of the interference patterns at $z = z_T/2$ for different values of the length δ_T. Such a cross-section is chosen because a self image of the slit grating appears phase-shifted by half period of the grating. For that reason we should see the interference fringes spaced between the slit sources of radiation.

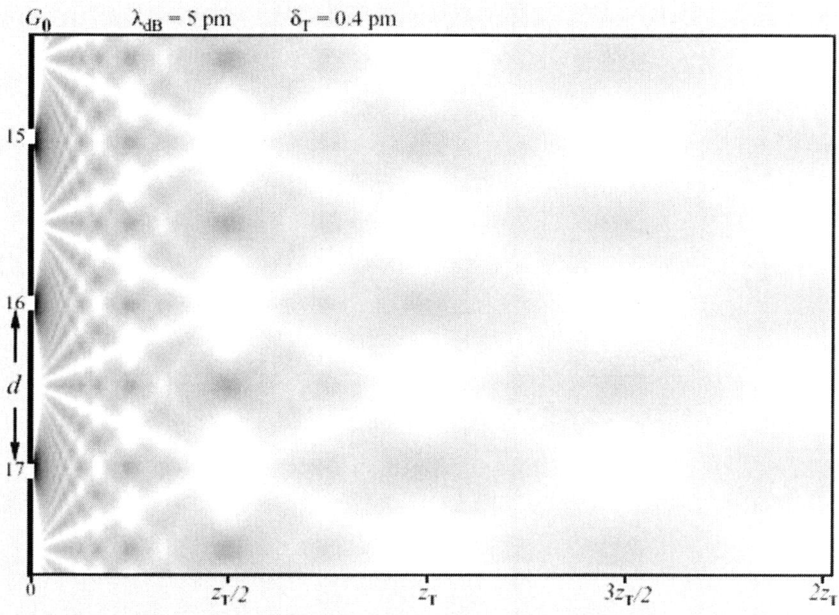

Figure 7. Blurred interference pattern in the near field. It is shown only the central part of the grating containing N_0 = 32 slits; d = 500 nm, δ_T = 0.4 pm < λ_{dB} = 5 pm.

Fig. 9 shows three characteristic patterns of the interference fringes. In Fig. 9(a) almost ideal interference fringes are shown obtained at δ_T < λ_{dB}. Fig. 9(b) shows interference fringes obtained at$\delta_T \sim \lambda_{dB}$. It is instructive to compare these interference fringes with those that have been measured in experiments [14, 34]. And Fig. 9(c) shows disappearance of interference fringes because of strong scattering of the particles on residual gases in vacuum, $\delta_T > \lambda_{dB}$.

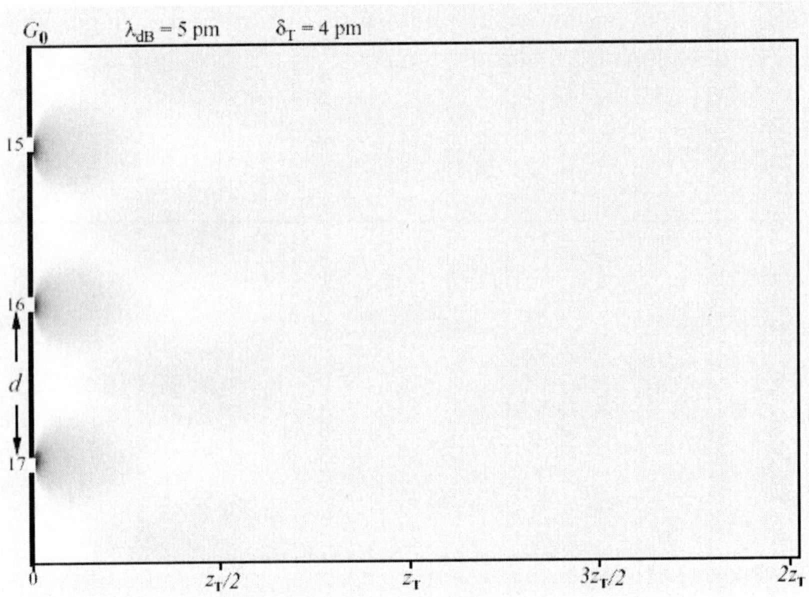

Figure 8. Destroyed interference pattern because of large $\delta_T = 4$ pm ~ $\lambda_{dB} = 5$ pm. $N_0 = 32$ slits, $d = 500$ nm.

Disappearance of interference fringes is numerically evaluated by calculating a characteristic called visibility [14, 34]. The fringe visibility [27] is represented as a ratio of difference between maximal and minimal intensities of the fringes to their sum:

$$V = \frac{P_{max} - P_{min}}{P_{max} + P_{min}}. \tag{40}$$

Evaluation of P_{max} and P_{min} is shown in Fig. 9(b). As follows from the figure, the evaluations are fulfilled in a central region of the grating. That is, edges of the grating have to be left far off from the measured zone. The visibility V as a function of the parameter δ_T is shown in Fig. 10. One can see that crossover from almost perfect interference fringes, $V = 1$, up to their absence, $V = 0$, begins near δ_T. ~λ_{dB}. Transition from almost coherent particle beam to incoherent is a cause of such a crossover [30].

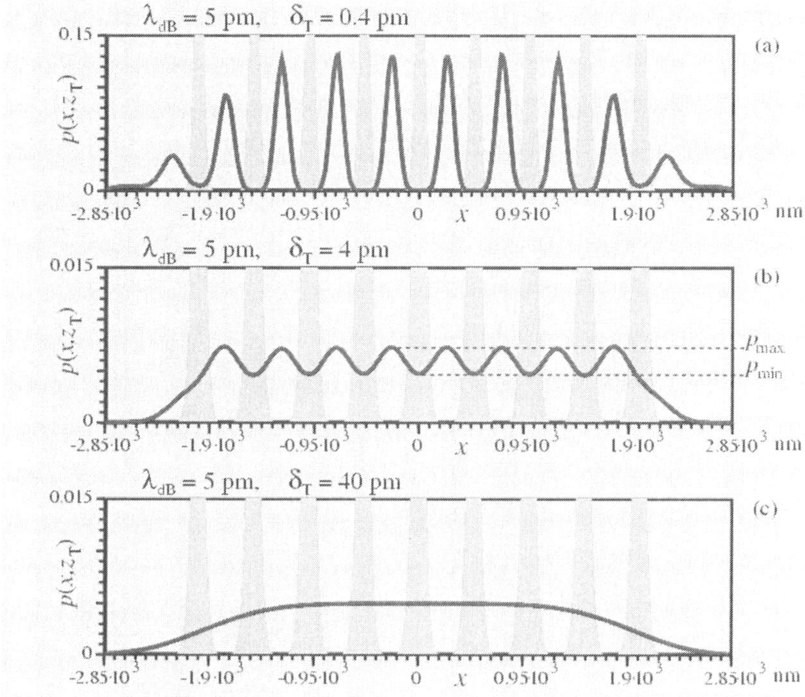

Figure 9. Interference fringes in cross-section of the density distribution pattern by the Talbot half-length, $z = z_T/2$, (the fringes are drawn in red): (a) δ_T = 0.4 pm, almost coherent beam; (b) δ_T = 4 pm, weak coherence; (c) δ_T = 40 pm, entirely noncoherent beam. Cyan strips show luminosity of slits. The grating consists of N_0 = 9 slits. Collapse of the interference pattern on edges of the grating is due to its finite length. Therefore visibility of the interference fringes is evaluated only for 5 central slits.

4. BOHMIAN TRAJECTORIES AND VARIANCE OF MOMENTA AND POSITIONS ALONG PATHS

Here we repeat computations of David Bohm [35] which lead to the Hamiltoton-Jacobi equation loaded by the quantum potential and, as consequence, to finding Bohmian trajectories. But instead of the

Schrödinger equation we choose Eq. (17) that contains complex-valued parameter $\Gamma=\beta+i\hbar$:

$$(\beta - i\hbar)\frac{\partial \psi}{\partial t} = \frac{(\beta^2 + \hbar^2)}{2m_N}\nabla^2\psi + \frac{(\beta - i\hbar)}{(\beta + i\hbar)}V\psi. \tag{41}$$

Here $\beta = 2k_B T\delta t$ (in particular, the diffusion coefficient reads $D = \beta/2mN$) and $V=U(\vec{q}_1)-TN$.

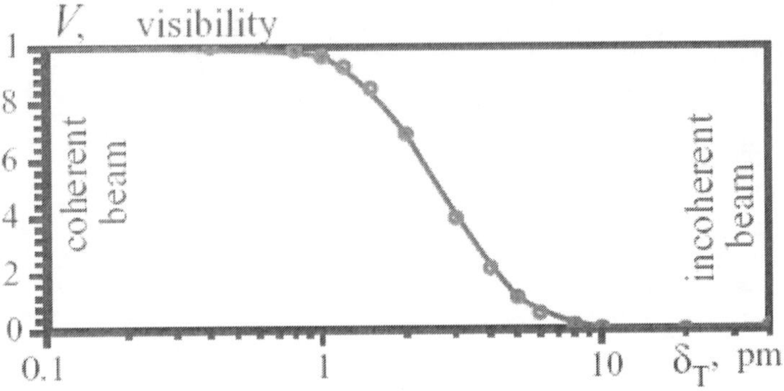

Figure 10. Visibility of interference fringes as a function of the parameter δ_T ranging from 0.1 to 40 pm. Wavelength of a matter wave is λ_{dB} = 5 pm.

Further we apply polar representation of the wave function, $\psi = R \exp\{iS/\hbar\}$. It leads to obtaining two equations for real and imaginary parts that deal with real-valued functions R and S. The function R is the amplitude of the wave function and $S/\hbar$ is its phase. After series of computations, aim of which is to put together real and imaginary terms, we obtain the following equations

$$\frac{\partial S}{\partial t} + \beta\frac{1}{R}\frac{\partial R}{\partial t} = -\underbrace{\frac{(\beta^2 + \hbar^2)}{2m_N\hbar^2}(\nabla S)^2}_{(a)} + \underbrace{\frac{(\beta^2 + \hbar^2)}{2m_N}\frac{\nabla^2 R}{R}}_{(b)} + \frac{(\beta^2 - \hbar^2)}{(\beta^2 + \hbar^2)}V, \tag{42}$$

$$\frac{\beta}{\hbar^2}\frac{\partial S}{\partial t} - \frac{1}{R}\frac{\partial R}{\partial t} = \frac{(\beta^2+\hbar^2)}{2m_N\hbar^2}\underbrace{(\nabla^2 S + 2\frac{1}{R}\nabla R\nabla S)}_{(c)} - \frac{2\beta}{(\beta^2+\hbar^2)}V. \tag{43}$$

Firstly, we can see that at $\beta = 0$ Eq. (42) reduces to the modified Hamilton-Jacobi equation due to loaded the quantum potential that is enveloped here by brace (b). And Eq. (43) reduces to the continuity equation. These equations read

$$\frac{\partial S}{\partial t} = -\underbrace{\frac{1}{2m_N}(\nabla S)^2}_{(a)} + \underbrace{\frac{\hbar^2}{2m_N}\frac{\nabla^2 R}{R}}_{(b)} + V, \tag{44}$$

$$-\frac{1}{R}\frac{\partial R}{\partial t} = \frac{1}{2m_N}\underbrace{(\nabla^2 S + 2\frac{1}{R}\nabla R\nabla S)}_{(c)}. \tag{45}$$

Terms enveloped by braces (a), (b), and (c) are the kinetic energy of the particle, the quantum potential, and the right part is a kernel of the continuity equation (45), respectively. In particular, the term $2\nabla R/R = \nabla \ln(R^2)$ relates to the osmotic velocity, see Eq. (23). Eqs. (44) and (45) are the same equations obtained by Bohm [35]. From historical viewpoint it should be noted that the same equations were published by Madelung[1] - in 1926 [36].

Momentum of the particle reads

$$\vec{p} = m_N\vec{v} = \nabla S, \tag{46}$$

where v⃗ v→ is its current velocity. The de Broglie equation relates the momentum p to the wavelength $\lambda_{dB} = h/p$, where $h = 2\pi\hbar$ is the Planck constant. Now, as soon as we found the current velocity

$$\vec{v} = \frac{1}{m_N}\nabla S = \frac{\hbar}{m_N}\mathrm{Im}\left(\left|\psi\right\rangle^{-1}\nabla\left|\psi\right\rangle\right) \tag{47}$$

positions of the particle in each current time beginning from the grating' slits up to a detector is calculated by the following formula

$$\vec{r}(t+\delta t)=\vec{r}(t)+\vec{v}(t)\delta t. \qquad (48)$$

Here t is a current time that starts from t=0 on a slit source and δt is an arbitrarily small increment of time. Some calculated trajectories of particles, the Bohmian trajectories, are shown in upper part ofFig. 5. It should be noted that the Bohmian trajectories follow from exact solutions of Eqs. (44)-(45). These equations give a rule for finding geodesic trajectories and secants of equal phases, $S/\hbar$, at given boundary conditions. The geodesic trajectories point to tendency of the particle migration along paths. And the secant surfaces describe a coherence of all the passing particles created on a single source.

In case of $\beta >> \hbar$ we have the following two equations

$$\frac{\partial R}{\partial t} = \underbrace{-\frac{\beta}{2m_N\hbar^2}R(\nabla S)^2}_{(a)} + \underbrace{\frac{\beta}{2m_N}\nabla^2 R}_{(b)}, \qquad (49)$$

$$\frac{\partial S}{\partial t} = \frac{\beta}{2m_N}\underbrace{(\nabla^2 S + 2\frac{1}{R}\nabla R\nabla S)}_{(c)}, \qquad (50)$$

Here we take into consideration that in the first equation we may replace $(\nabla S/\hbar)^2=4\pi^2/\lambda_{\mathrm{dB}}^2$. In the second equation we may replace the term ∇S∇S by mNv⃗ mNv→ as follows from Eq. (46). We notice also, that $2R^{-1}\nabla R=2\nabla\ln(R)=\nabla\ln(\rho)$. And $\beta/2m_N=D$ is the diffusion coefficient. Now we may rewriteEqs. (49)-(50) as follows

$$\frac{\partial R}{\partial t} = \underbrace{D\nabla^2 R}_{(b)} - \underbrace{\frac{4\pi^2}{\lambda_{\mathrm{dB}}^2}DR}_{(a)}, \qquad (51)$$

$$\frac{\partial S}{\partial t} = \underbrace{D\Big(\nabla^2 S + m_N(\nabla \ln(\rho), \vec{v})\Big)}_{(c)} = D\nabla^2 S + m_N(\vec{u}, \vec{v}). \tag{52}$$

Here $\vec{u} = D\,\nabla \ln(\rho)$is the osmotic velocity defined in Eq. (23). We got the two diffusion equations coupled with each other through sources. Namely, this coupling is provided due to the de Broglie wavelength and the osmotic velocity which can change with time. These diffusion equations cardinally differ from Eqs. (44)-(45). Because of diffusive nature of these supplementary parts blurring of interference patterns occurs. It leads to degeneration of the Bohmian trajectories to Brownian ones.

4.1. Dispersion Of Trajectories And The Uncertainty Principle

As for the Bohmian trajectories there is a problem concerning their possible existence. As follows fromEqs. (46) and (48) in each moment of time there are definite values of the momentum and the coordinate of a particle moving along the Bohmian trajectory. This statement enters in conflict with the uncertainty principle.

Here we try to retrace emergence of the uncertainty principle stemming from standard probability-theoretical computations of expectation value and variance of a particle momentum. We adopt a wave function in the polar representation

$$|\Psi\rangle = R\exp\{\mathrm{i}S/\hbar\}, \tag{53}$$

where $R = (\rho)^{1/2}$ is the amplitude of the wave function ($\rho = R^2 = \langle\Psi|\Psi\rangle$ is the probability density) and$S/\hbar$ is its phase. Momentum operator $\hat{p} = -i\hbar\nabla$and corresponding velocity operator $\hat{v} = -i(\hbar/m)\nabla$ $\hat{v} =$ $-i(\hbar/m)\nabla$are kinetic operators in quantum mechanics. Here m is mass of the particle. Expectation value of the velocity operator reads

$$\vec{V}_g = \frac{1}{\langle\Psi|\Psi\rangle}\langle\Psi|-\mathrm{i}\frac{\hbar}{m}\nabla|\Psi\rangle = \frac{1}{m}\left(\nabla S + \mathrm{i}\hbar\nabla S_Q\right). \tag{54}$$

The velocity V g is seen to be complex-valued. Here $S_Q = -\ln(R) = -\ln(\rho)/2$ is the quantum entropy [28] and $(\hbar/2m)$ is the quantum diffusion coefficient [6, 7]. Therefore its imaginary part is a quantum osmotic velocity

$$\vec{u}_Q = -(\hbar/m)\nabla S_Q = (\hbar/2m)\nabla\ln(\rho). \tag{55}$$

It is instructive to compare this velocity with the classical osmotic velocity given in Eq. (23). As can see the osmotic velocity stems from gradient of entropy that evaluates degree of order and disorder on a quantum level, likely of vacuum fluctuations.

Real part of Eq. (54) gives the current velocity v⃗ v→ defined by Eq. (47). It should be noted that because of existence of imaginary unit in definition of the momentum operator, real part of Eq. (54) is taken as the current velocity. Whereas imaginary unit is absent in Eq. (47). Therefore at computing the current velocity by Eq. (47) we take imaginary part.

Let us now calculate variance of the velocity V g. This computation reads

$$\mathrm{Var}(V_g) = \frac{1}{\langle\Psi|\Psi\rangle}\langle\Psi\left|\left(-\mathrm{i}\frac{\hbar}{m}\nabla - \vec{V}_g\right)^2\right|\Psi\rangle = \frac{1}{\langle\Psi|\Psi\rangle}\langle\Psi\left|\mathrm{i}\frac{\hbar}{m}\nabla\vec{V}_g + \underbrace{\mathrm{i}\frac{\hbar}{m}\vec{V}_g\nabla + V_g^2}_{(\mathrm{d})} - \frac{\hbar^2}{m^2}\Delta\right|\Psi\rangle. \tag{56}$$

Terms over bracket (d) kill each other as follows from Eq. (54). It is reasonable in the perspective to multiply Var(V g) by $m/2$

$$\frac{m}{2}\mathrm{Var}(V_g) = -\frac{1}{\langle\Psi|\Psi\rangle}\langle\Psi\left|\frac{\hbar^2}{2m}\Delta\right|\Psi\rangle + \mathrm{i}\frac{\hbar}{2}\nabla\vec{V}_g. \tag{57}$$

So, this expression has a dimensionality of energy. The first term to be computed represents the following result

$$-\frac{1}{\langle\Psi|\Psi\rangle}\left\langle\Psi\left|\frac{\hbar^2}{2m}\Delta\right|\Psi\right\rangle = \underbrace{\frac{1}{2m}(\nabla S)^2}_{(a)} - \underbrace{\frac{\hbar^2}{2m}\frac{\nabla^2 R}{R}}_{(b)} - \mathbf{i}\underbrace{\frac{\hbar}{2m}\left(\Delta S + 2\frac{(\nabla S, \nabla R)}{R}\right)}_{(c)}. \tag{58}$$

Here the term enveloped by bracket (a) is a kinetic energy of the particle, the term enveloped by bracket (b) with negative sign added is the quantum potential Q, and the term enveloped by bracket (c) comes from the continuity equation. See for comparison Eqs. (44) and (45). We rewrite the quantum potential as follows

$$Q = -\frac{\hbar^2}{2m}\frac{\nabla^2 R}{R} = -\frac{\hbar^2}{2m}(\nabla S_Q)^2 + \frac{\hbar^2}{2m}\nabla^2 S_Q. \tag{59}$$

As for the second term in Eq. (57) we have $i(\hbar/2)\nabla\vec{V}_g = i(\hbar/2m)\Delta S - (\hbar^2/2m)\Delta S_Q$.

. It is follows from computation by Eq. (54). As a result, the expression (57) takes the following view

$$\frac{m}{2}\mathrm{Var}(V_g) = \frac{1}{2m}(\nabla S)^2 - \frac{\hbar^2}{2m}(\nabla S_Q)^2 - \mathbf{i}\frac{\hbar}{m}\frac{(\nabla S, \nabla R)}{R}. \tag{60}$$

One can see that the variance consists of real and imaginary parts. Observe that the right side is represented through square of gradient of the complexified action [28], namely (∇(S+iħSQ))2/ 2m (∇(S+iħSQ))2/2m. We shall not consider here the imaginary part. Instead we shall consider the real part of this expression. It reads

$$\frac{m}{2}\mathrm{Re}(\mathrm{Var}(V_g)) = \frac{1}{2m}(\nabla S)^2 - \hbar\left(\frac{\hbar}{2m}(\nabla S_Q)^2\right). \tag{61}$$

The first term in this expression represents kinetic energy, E, of the particle. The second term, stemming from the quantum potential, contains under braces a term having dimensionality of inverse time, that is, of frequency

$$\omega_Q = \frac{\hbar}{2m}(\nabla S_Q)^2 \tag{62}$$

This frequency multiplied by $\hbar$ represents an energy binding a particle with vacuum fluctuations. This energy, as follows from Eq. (61), is equal to the particle mass multiplied by squared the osmotic velocity (55) and divided by 2. It is an osmotic kinetic energy. In the light of the above said we rewrite Eq. (61) in the following way

$$\frac{m}{2}\mathrm{Re}(\mathrm{Var}(V_g)) = E - \hbar\omega_Q \geq 0. \tag{63}$$

Let we have two Bohmian trajectories. Along one trajectory we have E_1 - $\hbar\omega Q_{,1}$, and along other trajectory we have a perturbed value E_2 - $\hbar\omega Q_{,2}$. Subtracting one from other we have

$$\delta E - \hbar\delta\omega_Q \geq 0. \tag{64}$$

One can suppose that emergence of the second trajectory was conditioned by a perturbation of the particle moving along the first trajectory. If it is so, then emergence of the second trajectory stems from an operation of measurement of some parameters of the particle. One can think that duration of the measurement is about $\delta t = 1/\delta\omega Q$. From here we find

$$\boxed{\delta E \delta t \geq \hbar} \tag{65}$$

Now let us return to Eq. (48) and rewrite it in the following view

$$\delta\vec{r}(t) = \vec{v}_1(t)\delta t \geq \vec{v}_1(t)\hbar/\delta E. \tag{66}$$

The initial Bohmian trajectory is marked here by subscript 1. Observe that $\delta E = m(v_2^2 - v_1^2)/2 \approx m\vec{v}_1\delta\vec{v}$. Here we have calculated $v_2^2 = (\vec{v}_1 + \delta\vec{v})^2 \approx v_1^2 + 2\vec{v}_1\delta\vec{v}$. Substituting computations of δE into Eq. (66) we obtain finally

$$\boxed{\delta\vec{p}\delta\vec{r} \geq \hbar} \tag{67}$$

Here we take into account $\delta\vec{p} = m\delta\vec{v}$.

5. CONCLUDING REMARKS

Each nanoparticle incident on a slit grating passes only through a single slit. Its path runs along a Bohmian trajectory which is represented as an optimal path for the nanoparticle migrating from a source to a detector. Unfortunately, the Bohmian trajectory can not be observable since a serious obstacle for the observation comes from the uncertainty principle. In other words, an attempt to measure any attribute of the nanoparticle, be it position or orientation, i.e., the particle momentum, leads to destroying information relating to future history of the nanoparticle. What is more, any collision of the nanoparticle with a foreign particle destroys the Bohmian trajectory which could give a real contribution to the interference pattern. It relates closely with quality of vacuum. In the case of a bad vacuum such collisions will occur frequently. They lead to destruction of the Bohmian trajectories. Actually, they degenerate to Brownian trajectories.

Excellent article [21] of Couder & Fort with droplets gives, however, a clear hint of what happens when the nanoparticle passes through a single grating slit. In the light of this hint we may admit that the particle "bouncing at moving through vacuum" generates a wave at each bounce. So, a holistic quantum mechanical object is the particle + wave. Here the wave to be generated by the particle plays a role of the pilot-wave first formulated by Lui de Broglie and later developed by Bohm [37]. It is interesting to note in this context, that the pilot-waves have many common with Huygens waves [5].

A particle passing through vacuum generates waves with wavelength that is inversely proportional to its momentum (it follows from the de Broglie formula, $\lambda = h/p$, where h is Planck's constant). One can guess that a role of the vacuum in the experiment of Couder & Fort takes upon itself a silicon oil surface with subcritical Faraday ripples activated on it [21]. Observe that pattern of the ripples is changed in the vicinity of extraneous bodies immersed in the oil which simulate grating slits. Interference of the ripples with waves generated by the bouncing droplets provides optimal paths for the droplets traveling through the slits and further. As a result we may observe an interference pattern emergent depending on amount of slits in the grating and distance between them.

Now we may suppose that the subcritical Faraday ripples on the silicon oil surface simulate vacuum fluctuations. Consequently, the vacuum fluctuations change their own pattern near the slit grating depending on amount of slits and distance between them. We may imagine that the particle passing through vacuum (bouncing through, Fig. 11) initiates waves which interfere with the vacuum fluctuations. As a result of such an interference the particle moves along an optimal path - along the Bohmian trajectory. Mathematically the bounce is imitated by an exponential term $\exp\{iS/\hbar\}$, where the angle $S/\hbar$ parametrizes the group of rotation given on a circle of unit radius. So, the path along which the particle moves is scaled by this unitary group, U(1), due to the exponential mapping of the phase $S/\hbar$ on the circle.

In conclusion it would be like to remember remarkable reflection of Paul Dirac. In 1933 Paul Dirac drew attention to a special role of the action S in quantum mechanics [38] - it can exhibit itself in expressions through $\exp\{iS/\hbar\}$. In 1945 he emphasized once again, that the classical and quantum mechanics have many general points of crossing [39]. In particular, he had written in this article: "We can use the formal probability to set up a quantum picture rather close to the classical picture in which the coordinates q of a dynamical system have definite values at any time. We take a number of times t_1, t_2, t_3, ... following closely one after another and set up the formal probability for the q's at each of these times lying within specified small ranges, this being permissible since the q's at any time all commutate. We then get a formal probability for the trajectory of the system in quantum mechanics lying within certain limits. This enables us to speak of some trajectories being improbable and others being likely."

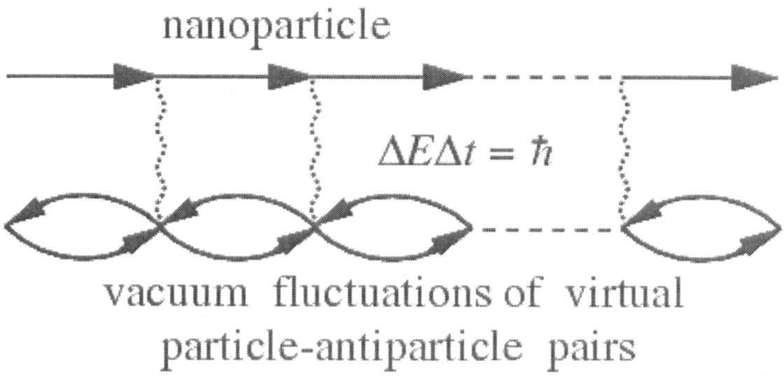

Figure 11. Bouncing a nanoparticle at moving through vacuum. Vertical dotted sinusoidal curves depict exchange by energy ΔE with vacuum virtual particle-antiparticle pairs over period of about $\Delta t = \hbar / \Delta E$.

ACKNOWLEDGEMENT

Author thanks Miss Pipa (administrator of Quantum Portal) for preparing programs that permitted to calculate and prepare Figures 5 to 8. Author thanks also O. A. Bykovsky for taking my attention to a single-particle interference observed for macroscopic objects by Couder and Fort and V. Lozovskiy for some remarks relating to the article.

NOTES

[1] - My attention to the Madelung' article was drawn by Prof. M. Berry.

REFERENCES

1. Dauben JW. Abraham Robinson: The creation of nonstandard analysis. A personal and mathematical odyssey. Princeton, NJ: Princeton University Press; 1995.

2. Stanford Encyclopeia of Philosophy: Newton's Views on Space, Time, and Motion. http://plato.stanford.edu/entries/newton-stm/ (accessed 12 August 2004).

3. Henson CW. Foundations of nonstandard analysis. In: Arkeryd L.O., Cutland N. J., & Henson C. W., (eds.) Nonstandard Analysis, Theory and application. The Netherlands: Kluwer Acad. Publ.; 1997,p. 1-51.

4. Nelson E. Radically Elementary Probability Theory. Princeton, New Jersey: Princeton Univ. Press; 1987.

5. Huygens C. Treatise on Light. Gutenberg eBook, No. 14725. http://www.gutenberg.org/files/14725/14725-h/14725-h.h (accessed 18 January 2005)

6. Nelson E. Dynamical theories of Brownian motion. Princeton, New Jersey: Princeton Univ. Press; 2001.

7. Nelson E. Quantum fluctuations. Princeton Series in Physics. Princeon, New Jersey: Princeton Univ. Press; 2001.

8. Feynman RP., Hibbs A. Quantum mechanics and path integrals. N. Y.: McGraw Hill; 1965.

9. Einstein A., von Smoluchowski M. Brownian motion. Moscow-Leningrad: ONTI (in Russian); 1936.

10. Sidorov LN., Troyanov SI. At the dawn of a new chemistry of fullerenes. PRIRODA 2011; 1153(9) 22-30.

11. Arndt M., Hackermüller L., Reiger E. Interferometry with large molecules: Exploration of coherence, decoherence and novel beam methods. Brazilian Journal of Physics 2005; 35(2A) 216-223.

12. Brezger B., Arndt M., Zeilinger A. Concepts for near-field interferometers with large molecules. J. Opt. B: Quantum Semiclass. Opt. 2003; 5(2) S82-S89.

13. Brezger B., Hackermüller L., Uttenthaler S., Petschinka J., Arndt M., Zeilinger A. Matter-Wave Interferometer for Large Molecules. Phys. Rev. Lett. 2002; 88 100404.

14. Gerlich S., Eibenberger S., Tomandl M., Nimmrichter S., Hornberger K., Fagan P. J., Tuxen J., Mayor M., Arndt, M. Quantum interference of large organic molecules. Nature Communications 2011; (2) 263. http://www.nature.com/ncomms/journal/v2/n4/full/ncomms1263.html (accessed 5 April 2011)

15. Hornberger K., Hackermüller L., Arndt M. Inuence of molecular temperature on the coherence of fullerenes in a near-field interferometer. Phys. Rev. A. 2005; 71, 023601.

16. Hornberger K., Sipe JP., Arndt M. Theory of decoherence in a matter wave Talbot-Lau interferometer. Phys. Rev. A. 2004; 70, 053608.

17. Nimmrichter S., Hornberger K. Theory of near-fieldmatter wave interference beyond the eikonal approximation. Phys. Rev. A. 2008; 78, 023612.

18. Juffmann,T., Truppe S., Geyer P., Major AG., Deachapunya S., Ulbricht H., Arndt, M. Wave and particle in molecular interference lithography. Phys. Rev. Lett. 2009; 103, 263601.

19. Berry M., Klein S. Integer, fractional and fractal Talbot effects, Journal of Modern Optics, 1996; 43(10) 2139-2164.

20. Berry M., Marzoli L., Schleich W. Quantum carpets, carpets of light. Physics World, 2001; (6), 39-44.

21. Couder Y., Fort E. Single-Particle Diffraction and Interference at a Macroscopic Scale. Phys. Rev. Lett. 2006; 97 (6), 154101(4).

22. Stratonovich RL. Topics in the theory of random noise. N.Y.: Gordon and Breach; 1963.

23. Gnedenko BV. Theory of Probability. N.Y.: Gordon & Breach; 1997.

24. Ventzel AD. The course of stochastic processes theory. Moskow: Nauka (in Russian); 1975.

25. Grosche C. An introduction into the Feynman path integral. http://arXiv.org/abs/hep-th/9302097 (accessed 20 February 1993).

26. MacKenzie R. Path integral methods and applications. http://arXiv.org/abs/quant-ph/0004090 (accessed 20 April 2000).

27. Juffmann T., Milic A., Müllneritsch M., Asenbaum P., Tsukernik A., Tüxen J., Mayor M., Cheshnovsky O., Arndt, M. Real-time single-molecule imaging of quantum interference. Nature Nanotechnology - Letter. 2012; 7, 297–300. http://www.nature.com/nnano/journal/v7/n5/full/nnano.2012.34.html?WT.ec_id=NNANO-201205 (accessed 25 March 2012)

28. Sbitnev VI. Bohmian trajectories and the path integral paradigm - complexified Lagrangian mechanics. In: Pahlavani MR. (ed.) Theoretical Concepts of Quantum Mechanics. Rijeka: InTech; 2011 p313-340. http://www.intechopen.com/books/theoretical-concepts-of-quantum-mechanics/the-path-integral-paradigm-and-bohmian-trajectories-from-the-lagrangian-mechanics-to-complexified-on (accessed 2004-2012)

29. Nairz O., Arndt M., Zeilinger A. Quantum interference experiments with large molecules. Am. J. Phys. 2003; 71 (4), 319-325.

30. Sbitnev VI. Matter waves in the Talbot-Lau interferometry, http://arxiv.org/abs/1005.0890 (accessed 17 September 2010).

31. Volokitin AI. Quantum drag and graphene. PRIRODA, 2011; 1153(9), 13-21.

32. Mandel L., Wolf E. Optical coherence and quantum optics. Cambridge: Cambridge University Press; 1995.

33. McMorran B.,Cronin AD. Model for partial coherence and wavefront curvature in grating interferometers. Phys. Rev. A. 2008; 78 (1), 013601(10).

34. Gerlich S., Hackermüller L., Hornberger K., Stibor A., Ulbricht H., Gring M., Goldfarb F., Savas T., Müri M., Mayor M., Arndt, M. A Kapitza-Dirac-Talbot-Lau interferometer for highly polarizable molecules. Nature Physics. 2007; 3, 711-715.

35. Bohm D. A suggested interpretation of the quantum theory in terms of "hidden variables", I & II. Physical Review, 1952; 85, 166-179 & 180-193.

36. Madelung E. Quantentheorie in Hydrodynamischer form. Zts. f. Phys. 1926; 40, 322-326.

37. Stanford Encyclopeia of Philosophy: Bohmian Mechanics. http://plato.stanford.edu/entries/qm-bohm/ (accessed 26 October 2001).

38. Dirac PAM. The Lagrangian in Quantum Mechanics. Physikalische Zeitschrift der Sowjetunion. 1933; 3, 64-72.

39. Dirac PAM. On the analogy between classical and quantum mechanics. Rev. Mod. Phys. 1945; 17 (2 & 3), 195-199.

7

DISTRIBUTIONAL APPROACH TO POINT INTERACTIONS IN ONE-DIMENSIONAL QUANTUM MECHANICS

Marcos Calçada[1], José T. Lunardi[1*], Luiz A. Manzoni[2] and Wagner Monteiro[1]

[1]Grupo de Física Teórica e Modelagem Matemática, Departamento de Matemática e Estatística, Universidade Estadual de Ponta Grossa, Ponta Grossa, PR, Brazil

[2]Department of Physics, Concordia College, Moorhead, MN, USA

1. INTRODUCTION

Point (zero-range) interactions have attracted great interest in quantum mechanics [1–7]. They provide important solvable models with a wide variety of applications in atomic physics, such as the Lieb-Liniger [8, 9] model for a one dimensional gas of bosons interacting by means of a δ-function potential (for other applications see e.g., [10–12] and references therein). In addition, point interactions have proven to be a fruitful *theoretical laboratory* to investigate methods of quantum field theory (QFT), such as regularization and renormalization, in simpler and more manageable models in one and two-dimensional quantum mechanics [13–15].

The simplest point interaction in one dimensional non-relativistic quantum mechanics is given by the Dirac's δ-function potential, which is

well defined and has well-known solutions. However, attempts to consider more general interactions, such as that associated with a δ' potential (the prime indicates a spatial derivative), have been known to be plagued with difficulties associated with the definition of the interaction [6, 16–21] (also see [22] and references therein).

Mathematically rigorous methods were employed to define such interactions by obtaining self-adjoint extensions (SAE) of the kinetic energy operator for Schrödinger's theory. It was shown that a four-parameter family of interactions, defined by their boundary conditions, exhaust all the possibilities for point interactions in non-relativistic quantum mechanics [6, 23, 24] (for a review of the mathematical literature on this subject see [1]).

Given the somewhat abstract and mathematically involved nature of SAE, another approach was developed for this subject, consisting in the investigation of regularizations using sequences of regular short-range potentials which converge to point potentials in the zero-range limit [16, 20, 21, 25]. Even if this method is more intuitive and appealing from a physicist's point of view, such an approach has often led to ambiguous and even contradictory results—which, in general, arise from the dependence of the particular regularizing scheme employed (see, for instance, [5, 7, 26] and references there cited).

In relativistic quantum mechanics, for a Dirac particle, even the δ-function potential (which is the most singular interaction allowed in the Dirac's equation) is problematic and it is known to lead to ambiguities [27–30]. Similarly to the non-relativistic case, the SAE approach implies that the most general point interactions in Dirac's theory are also given by a four-parameter family of interactions [31, 32]. The most commonly studied particular cases of relativistic point interactions correspond to pure electrostatic and pure scalar potentials, which have also been investigated by regularization through δ-converging sequences of functions, with results that are, in general, dependent on the regularizing function used [5,27, 28].

The situation described above is not entirely new: analogous difficulties involving regularization ambiguities arise in QFT and have been successfully addressed by using the Epstein-Glaser approach to QFT, in which distribution theory is employed from the beginning and the symmetries of the system are carefully implemented (see e.g., [33–36] and references therein). In a similar vein, in this paper we will introduce a distributional approach to point interactions in quantum mechanics – first announced in a preliminary version in [37].

The method proposed here starts from the realization that both the singular potential and the wave function in the Schrödinger or Dirac equations *must* be considered as distributions. This precludes the naive definition of the interaction term in these equations as the usual product between the wave function and the potential energy, since such a product is generally ill-defined in distribution theory [38, 39]. We will introduce a well defined distribution to play the role of the interaction term, the properties of which must follow from simple mathematical requirements concerning its support and singular order (a concept introduced in the next section), besides the physical requirement of conservation of the probability current across the singular point. As a consequence, a four-parameter family of distributions will naturally emerge as the most general point interaction in both non-relativistic and relativistic one-dimensional quantum mechanics, a result which is in complete agreement with those obtained with the SAE approach [32, 40]. However, there is an important conceptual difference between the SAE and the distributional approaches, for while the first characterizes all the possible point interactions only through the boundary conditions that the wave function must satisfy around the singular point, the latter explicitly gives the interaction term as a distribution concentrated at that point—thus, in the distributional approach the boundary conditions emerge as a *consequence* of the properties of the particular interaction distribution substituted into the Schrödinger or Dirac equations. Thus, the distributional approach provides an alternative (also mathematically rigorous) to the method of SAE, with the advantage of being less abstract (to a physicist) and allowing a more intuitive physical interpretation of the interaction term, which here is presented explicitly as a distribution. In addition, the distributional approach can contribute to clarify some ambiguities which arise from the use of regularizations to treat zero-range potentials – in particular, the explicit form of the interaction distribution allow us to address, in a straightforward way, the much debated question about the existence or not of a "true" (i.e., *odd* under parity) point interaction associated with a δ' potential. Notice that a proper definition of the notion of an odd point interaction is a non-trivial task in the context of the SAE approach.

It is worth to notice that in this work we are concerned only to the well-established Schwartz's theory of distributions, which considers only distributions defined on spaces of infinitely differentiable test functions. An approach based on an alternative theory of distributions defined on spaces of discontinuous test functions was also proposed by Kurasov [41], who assumes the wave function to be a member of a space of test functions and claims that point interactions cannot be properly defined in the framework of the standard distribution theory. In the approach proposed here such a difficulty does not arise because the wave function itself must be consistently treated as a distribution, not as a test function. This is analogous to what occurs in the Epstein-Glaser distributional approach to QFT, in which quantities defined in terms of quantum fields must be treated themselves as (operator-valued) distributions [33, 34].

This paper is organized as follows. In section 2 we introduce the distributional approach by considering the Schrödinger equation with a singular potential of order $r_s = 1$, the most singular point interaction allowed in non-relativistic one-dimensional quantum mechanics. In Section 3 we extend our methods to consider Dirac's equation with a singular interaction of order $r_s = 0$. Section 4 is reserved for our concluding remarks.

2. SCHRÖDINGER'S EQUATION WITH POINT INTERACTIONS

In this Section we introduce the distributional approach to point interactions in one dimensional quantum mechanics. We consider distributions defined on the Schwartz space of rapid descent test functions—for a comprehensive treatment of distributions see [38], which we follow closely (also see [39]).

For a non-relativistic particle moving under the influence of a potential energy $V(x)$ which is assumed to be a locally integrable function (i.e., a function integrable in the Lebesgue sense over every finite interval), the one-dimensional time-independent Schrödinger equation can be written as (in this section we will adopt units such that $\hbar = 1$ and $2m = 1$)

$$\frac{d^2}{dx^2}\psi(x) + E\psi(x) = V(x)\psi(x)\,. \tag{1}$$

Then, by assuming that the wave function ψ is continuous everywhere, the *interaction term*, $V(x)\psi(x)$, is well-defined and corresponds to a locally integrable ordinary function. On the other hand, potentials which do not correspond to locally integrable functions are called *singular*, as it is the case of the Dirac δ-"function" or its derivatives, and only make sense as distributions. Distributions corresponding to locally integrable functions are said to be *regular*. As a consequence, in the case of singular potentials the Schrödinger equation is in general mathematically ill-defined, since the wave function itself must be considered as a distribution and the product appearing on the r. h. s. of Equation (1) is not a well defined operation for two arbitrary distributions [38, 39].

The starting point of a distributional approach to singular interactions is to rewrite Schrödinger's equation as

$$\psi''(x) + k^2\psi(x) = s[\psi](x)\,, \tag{2}$$

where $k \equiv \sqrt{E}$, the prime indicates a distributional derivative with respect to x, and the interaction term was substituted by a distribution $s[\psi](x)$, still to be determined by mathematical and physical requirements, and which univocally determines the interaction. Then, in any interval which does not include the singularities the potential is regular (and the wave function ψ continuous), and therefore the interaction distribution, $s[\psi](x)$, coincides with the ordinary product of functions $\psi(x)V(x)$ in such an interval.

In order to determine the interaction distribution $s[\psi](x)$ let us define the *singular order* (or simply the*order*) of a distribution, a concept that characterizes the "strength" of the singularity and plays a crucial role in this approach. The definition of singular order of a distribution adopted in this work is the extended concept of order considered by Zemanian (see [38], p. 162–163), which can be written as follows. A distribution f is said to have singular order $r \in \mathbb{Z}$ on a closed finite interval I if $f = v^{(r+2)}$ in this interval, where v is a distribution corresponding to a continuous but not differentiable function on I and $v^{(r+2)}$ is the $(r + 2)$-th distributional derivative of v (a "derivative of negative order" means an indefinite integration, as usual). An infinitely smooth function is said to have order $r = -\infty$. If a distribution has a finite order the operations of differentiation and indefinite integration change its order by +1 and −1, respectively. It follows that distributions having $r \leq -2$ are regular, whereas distributions having $r \geq 0$ are singular. When $r = -1$ the distribution can be regular or singular, depending on the case. As some important examples, the Heaviside θ distribution has singular order −1 (and it is a regular distribution), while its primitives have $r = -2$, and correspond to continuous but not differentiable functions. The Dirac's delta δ $(= \theta')$ and its first derivative δ' are singular distributions with singular orders 0 and +1, respectively.

Let us now return to the task of determining the interaction distribution $s[\psi](x)$. We require that:

(i) supp $s \subseteq$ supp V;

(ii) the singular order of the distribution $s[\psi](x)$ cannot exceed the singular order of $V(x)$ in any closed subset of its support,

where supp s indicates the *support* of the distribution s, which is the smallest closed set outside of which $s$equals zero [38]. Both these requirements are automatically satisfied when $V(x)$ corresponds to a regular distribution—here we are just extending them in a natural way to the case of potentials described by singular distributions. In particular, requirement (*ii*) prevents the introduction of interaction terms even more singular than the original potential (this is analogous to the *minimal distribution splitting* condition introduced in the Epstein-Glaser approach to QFT [33, 34]).

In this paper we are interested only in singular *point interactions*, i.e., interactions which vanish everywhere except at a single point, which is assumed to be the origin $x = 0$, where they are singular. Thus, from the requirement (*i*) above, it follows that supp $s[\psi] = \{0\}$.

We now recall a fundamental theorem from distribution theory (see [38], Theorem 3.5-2, p. 98):

Theorem 1. A necessary and sufficient condition for a distribution $f(x)$ on $\mathbb{R}$ to have a support concentrated at the origin is that it be a finite sum

$$f(x) = \sum_{m=0}^{r_s} \alpha_m \, \delta^{(m)}(x) \, , \tag{3}$$

where the α_m are (complex) constants, $\alpha_{r_s} \neq 0$, $\delta^{(m)}$ is the m-th derivative of the δ distribution, and r_s is the singular order of distribution f.
From this theorem we conclude that the interaction distribution $s[\psi](x)$ must be a linear combination of the delta distribution and its derivatives at the origin, which implies

$$s[\psi](x) = \sum_{m=0}^{r_s} \alpha_m[\psi] \, \delta^{(m)}(x) \, , \tag{4}$$

where r_s stands for the *singular order* of the interaction term, $\delta^{(m)}$ denotes the m-th derivative of the δ-distribution and $\alpha_m[\psi]$ are complex numbers, expressed as *linear functionals* depending on the behavior of the wave function *around the singular point,* i.e., $\alpha_m[\psi] \equiv \alpha_m[\psi(0^\pm), \psi'(0^\pm)]$. This last condition is needed if only *local* interactions are to be considered, and the linearity of $\alpha_m[\psi]$ is required by the superposition principle.

The specification of the singular order r_s and of all the coefficients α_m ($m = 0, \ldots, r_s$) in Equation (4) uniquely determines the interaction. However, in order to determine such quantities we must impose on the system some fundamental physical requirements and, since we are considering a *stationary system*, it is natural to require that

(iii) the probability current must be conserved everywhere – in particular, it must be conserved across the singular point;

there is, of course, nothing new about this requirement in itself – however, when considered together with (*i*) and (*ii*) it drastically reduces the number of free parameters in a point interaction, as shown below.

From now on, let us restrict ourselves to point interactions of singular order $r_s = 1$ and look for the most general $s[\psi](x)$ allowed by the requirements (*i*)-(*iii*). We observe that $r_s = 1$ is the maximum singular order allowed in order to have normalizable wave functions (i.e., square integrable ordinary functions)—if $r_s > 1$ were allowed the stationary wave function would have singular order greater than or equal to zero, and it would necessarily be a singular distribution.

An example of a potential with singular order $r_s = 1$ is given by the so-called delta prime potential energy, namely $V(x) = \gamma\delta'(x)$ with the strength of the interaction γ being a real constant. This potential, when naively substituted in Schrödinger's Equation (1), is well-known to lead to contradictions—see, e.g., [6, 19,22]. The reason for such contradictions is that the product $\psi(x)\delta'(x)$ can be defined in distribution theory if, *and only if*, both the wave function *and* its first derivative *are continuous at the origin* (in which case it is $\psi(x)\delta'(x) = -\psi'(0)\delta(x) + \psi(0)\delta'(x)$, see [38]) – but such requirement is not compatible with Schrödinger's Equation (1), as a simple analysis of the singular order shows: both sides of this equation must have the same singular order at the origin, and the r.h.s. $-\psi'(0)\delta(x) + \psi(0)\delta'(x)$ has singular order +1; hence $\psi''(x)$, $\psi'(x)$ and $\psi(x)$ must have singular orders +1, 0, and −1, respectively, which means that*the wave function must be discontinuous* at the origin.

Let us return to our task of determining the $s[\psi](x)$ which corresponds to a point interaction with singular order $r_s = +1$. According to conditions (*i*), (*ii*) and Theorem 1, one must consider (2) for the most general interaction distribution with singular order +1 and concentrated at the origin:

$$s\left[\psi\right](x) = \alpha_0\left[\psi\right]\delta(x) + \alpha_1\left[\psi\right]\delta'(x), \tag{5}$$

with $\alpha_1\ [\psi] \neq 0$. The coefficients $\alpha_0\ [\psi]$ and $\alpha_1[\psi]$ have yet to be determined (the δ-function potential, which has singularity $r_\delta = 0$, can be obtained as a particular case of the above interaction by requiring $\alpha_0[\psi] \neq 0$ and $\alpha_1\ [\psi] = 0$).

We recall that taking the indefinite integral of a distribution decreases its singular order by one. Thus, since ψ must have singular order −1, any primitive $\psi^{(-1)}$ must have singular order −2, and corresponds to a function continuous (but not differentiable) at the origin. Therefore, by taking the indefinite integral in both sides of Equation (2), with the interaction distribution given by Equation (5), we have

$$\psi'(x) + k^2\psi^{(-1)}(x) = \alpha_0\left[\psi\right]\theta(x) + \alpha_1\left[\psi\right]\delta(x) + c_1, \tag{6}$$

where c_1 is an arbitrary (complex) constant. In any interval which does not include the origin both sides of the above equation equal an ordinary function (in particular, the distribution δ vanishes identically in such an interval). Taking into account the continuity of $\psi^{(-1)}$ at the origin, from the above expression we have

$$\psi'\left(0^+\right) - \psi'\left(0^-\right) = \alpha_0\left[\psi\right]. \tag{7}$$

Similarly, taking the indefinite integral of Equation (6), and using the continuity of any primitive $\theta^{(-1)}$ of the distribution θ (since such a primitive has singular order −2), we conclude that

$$\psi\left(0^+\right) - \psi\left(0^-\right) = \alpha_1\left[\psi\right]. \tag{8}$$

Equations (7, 8) reflect the fact that the boundary conditions (b.c.) at the origin are completely determined from the knowledge of the interaction, i.e., from the functional coefficients α_0 and α_1.

The above b.c. can be written in terms of the wave function and its derivative by noticing that the most general way to express the linear functionals α_0 and α_1 is

$$\Omega \equiv \begin{pmatrix} \alpha_1 \\ \alpha_0 \end{pmatrix} = M_+ \Phi\left(0^+\right) - M_- \Phi\left(0^-\right), \qquad \text{with} \qquad \Phi(x) \equiv \begin{pmatrix} \psi(x) \\ \psi'(x) \end{pmatrix} \tag{9}$$

where $M_\pm$ are 2×2 complex matrices yet to be determined.

Equation (9) is a direct consequence of requirements (*i*) and (*ii*) and the solution of the *distributional*Schrödinger equation with the general interaction (5). This alone, however, does not impose any constraints on the form of $M_\pm$ (i.e., on the functionals α_0 and α_1); such constraints will come from imposing, in addition to (*i*) and (*ii*), condition (*iii*).

Conservation of the probability current across the origin simply means $j(0^-) = j(0^+)$ and, since in any interval which does not include the origin the distribution describing the probability current coincides with the ordinary product of functions $j(x) = -i\left[\psi^*(x)\psi'(x) - \psi^{*\prime}(x)\psi(x)\right]$, both sides of that equation are well defined and finite for a general point interaction.

In order to impose (*iii*) it is convenient first to use the identity $|z+iw|^2 - |-z+iw|^2 = \frac{2}{i}\left(zw^* - z^*w\right)$, which is valid for any pair of complex numbers *z* and *w*, and rewrite the current density as [42]

$$L_0 j(x) = \frac{1}{2}\left[|L_0\psi'(x) + i\psi(x)|^2 - |-L_0\psi'(x) + i\psi(x)|^2\right], \tag{10}$$

where L_0 is an arbitrary non-zero constant inserted for dimensional reasons. Introducing the vectors

$$V_1[\psi] \equiv \begin{pmatrix} L_0\psi'(0^+) + i\psi(0^+) \\ -L_0\psi'(0^-) + i\psi(0^-) \end{pmatrix} ; \quad V_2[\psi] \equiv \begin{pmatrix} -L_0\psi'(0^+) + i\psi(0^+) \\ L_0\psi'(0^-) + i\psi(0^-) \end{pmatrix} \tag{11}$$

the condition of probability current conservation, $j(0^+) = j(0^-)$, can be rewritten as the requirement that the vectors $V_1[\psi]$ and $V_2[\psi]$ have the same length: $V^\dagger_2[\psi]V_2[\psi] = V^\dagger_1[\psi]V_1[\psi]$. Thus, there must exist an unitary matrix U such that [42]

$$V_2[\psi] = UV_1[\psi] \tag{12}$$

The boundary conditions (7–9) can be written in matrix form in terms of the vectors $V_1[\psi]$ and $V_2[\psi]$ as

$$\Omega = A_1V_1[\psi] - A_2V_2[\psi] \,, \tag{13}$$

With

$$A_1 = \frac{1}{2}\begin{pmatrix} -i & i \\ L_0^{-1} & L_0^{-1} \end{pmatrix} \,, \quad A_2 = \frac{1}{2}\begin{pmatrix} i & -i \\ L_0^{-1} & L_0^{-1} \end{pmatrix} , \tag{14}$$

and after using the condition of probability current conservation, Equation (12), the above expression for Ω becomes

$$\begin{aligned} \Omega &= [A_1 - A_2U]\, V_1[\psi] \\ &= \frac{1}{2}\begin{pmatrix} -i\left[1+e^{i\theta}(z+w^*)\right] & i\left[1+e^{i\theta}(z^*-w)\right] \\ L_0^{-1}\left[1-e^{i\theta}(z-w^*)\right] & L_0^{-1}\left[1-e^{i\theta}(z^*+w)\right] \end{pmatrix} V_1[\psi]. \end{aligned} \tag{15}$$

where we used the following parametrization for the unitary matrix U:

$$U = e^{i\theta}\begin{pmatrix} z & w \\ -w^* & z^* \end{pmatrix}, \quad \theta \in [0,\pi), \quad |z|^2 + |w|^2 = 1. \tag{16}$$

After rewriting $V_1[\psi]$ in (15) in terms of $\Phi(0^\pm)$ we obtain Ω in the form (9), with $M_\pm$ given by

$$M_{+} = \frac{1}{2}\begin{pmatrix} 1+e^{i\theta}(z+w^{*}) & -iL_0\left[1+e^{i\theta}(z+w^{*})\right] \\ iL_0^{-1}\left[1-e^{i\theta}(z-w^{*})\right] & 1-e^{i\theta}(z-w^{*}) \end{pmatrix}$$

$$M_{-} = \frac{-1}{2}\begin{pmatrix} -1-e^{i\theta}(z^{*}-w) & -iL_0\left[1+e^{i\theta}(z^{*}-w)\right] \\ iL_0^{-1}\left[1-e^{i\theta}(z^{*}+w)\right] & -1+e^{i\theta}(z^{*}+w) \end{pmatrix} \quad (17)$$

then, the b.c. (13) can be rewritten as

$$R_{+}\Phi\left(0^{+}\right) = R_{-}\Phi\left(0^{-}\right), \quad (18)$$

where we defined $R_{\pm} = 1 - M_{\pm}$. It is easy to show that det $R_{+} = -w^{*}e^{i\theta}$ and that det $R_{-} = we^{i\theta}$. Thus, R_{+} and R_{-} are *both* invertible if, and only if, $w \neq 0$, in which case we can write

$$\Phi\left(0^{+}\right) = \Lambda\,\Phi\left(0^{-}\right), \quad (19)$$

where

$$\Lambda = R_{+}^{-1}R_{-} = \frac{i}{w^{*}}\begin{pmatrix} \sin\theta - \Im(z) & L_0(\cos\theta + \Re(z)) \\ \frac{-\cos\theta + \Re(z)}{L_0} & \sin\theta + \Im(z) \end{pmatrix}, \quad (20)$$

with $\Re(z)$ and $\Im(z)$ indicating the real and imaginary parts of z, respectively. From (20), it follows that $|\det\Lambda| = |-\frac{w}{w^{*}}| = 1$ and Λ can be rewritten in the form

$$\Lambda = e^{i\varphi}\begin{pmatrix} a & b \\ c & d \end{pmatrix}, \quad ad - bc = 1, \quad \varphi \in [0,\pi) \quad \text{and} \quad a,b,c,d \in \mathbb{R}. \quad (21)$$

The four parameter family of point interactions given by Equation (21) represents the most general point interactions consistent with a rigorous distributional approach to Schrödinger equation, under the requirements (*i*)-(*iii*) and the condition det $R_{\pm} \neq 0$ [see (18)]. This result is in agreement with the results obtained by the method of SAE, and corresponds to the case of a four parameter family of *non-separated*solutions [40].

From Equations (13, 19) in the case $w \neq 0$ one can rewrite the functionals α_1 and α_0 in terms of $\Phi(0^-)$ as

$$\Omega = \Phi(0^+) - \Phi(0^-) = (\Lambda - 1)\,\Phi\left(0^-\right)\ , \tag{22}$$

from which one can obtain the explicit form of the interaction distribution as

$$\begin{aligned} s\,[\psi]\,(x) &= \left[c\,e^{i\varphi}\psi\left(0^-\right) + \left(d\,e^{i\varphi} - 1\right)\psi'\left(0^-\right)\right]\delta(x) \\ &+ \left[\left(a\,e^{i\varphi} - 1\right)\psi\left(0^-\right) + b\,e^{i\varphi}\psi'\left(0^-\right)\right]\delta'(x)\,. \end{aligned} \tag{23}$$

Alternatively, one can invert (19) and substitute into (22) to obtain

$$\begin{aligned} s\,[\psi]\,(x) &= \left[c\,e^{-i\varphi}\psi\left(0^+\right) - \left(a\,e^{-i\varphi} - 1\right)\psi'\left(0^+\right)\right]\delta(x) \\ &- \left[\left(d\,e^{-i\varphi} - 1\right)\psi\left(0^+\right) - b\,e^{-i\varphi}\psi'\left(0^+\right)\right]\delta'(x)\,. \end{aligned} \tag{24}$$

In order to complete the analysis of Schrödinger's equation with point interactions, we still have to consider the case in which $w = 0$ (which reduces the number of free parameters to 2). In this case both $\det R_\pm = 0$ and, from Equation (12), we obtain

$$(ze^{i\theta} - 1)\psi(0^+) - iL_0(ze^{i\theta} + 1)\psi'(0^+) \;=\; 0, \tag{25}$$

$$(z^*e^{i\theta} - 1)\psi(0^-) + iL_0(z^*e^{i\theta} + 1)\psi'(0^-) \;=\; 0. \tag{26}$$

The solution of these equations is as follows. If $ze^{i\theta} + 1 = 0$ $(z^*e^{i\theta} + 1 = 0)$ we obtain $\psi(0^+) = 0$ with $\psi'(0^+)$ an arbitrary finite number [$\psi(0^-) = 0$ and $\psi'(0^-)$ arbitrary, respectively]. On the other hand, if $ze^{i\theta} + 1 \neq 0$ it follows that $\psi'(0^+) = h_+\,\psi\,(0^+)$, with $h_+ = \frac{ze^{i\theta}-1}{iL_0(ze^{i\theta}+1)} = \frac{2\Im(ze^{i\theta})}{L_0|1+ze^{i\theta}|^2}$ a real number [analogously, if $z^*e^{i\theta} + 1 \neq 0$, we have $\psi'(0^-) = h_-\,\psi\,(0^-)$ with

$h_{+} = \frac{ze^{i\theta}-1}{iL_0(ze^{i\theta}+1)} = \frac{2\Im(ze^{i\theta})}{L_0|1+ze^{i\theta}|^2}$ real]. Therefore, one can summarize the case $w = 0$ as

$$\psi'\left(0^{\pm}\right) = h^{\pm}\psi\left(0^{\pm}\right), \quad h^{\pm} \in \mathbb{R} \cup \{\infty\}, \tag{27}$$

which, for instance, implies that $j(0^+) = j(0^-) = 0$. Again, this result is in agreement with the results obtained from the SAE approach, and corresponds to a two-parameter family of *separated* solutions [40]. In this case the explicit form of the interaction distribution is

$$\begin{aligned} s[\psi](x) &= \left[h_+\psi(0^+) - h_-\psi(0^-)\right]\delta(x) + \left[\psi(0^+) - \psi(0^-)\right]\delta'(x) \\ &= \left[\psi'(0^+) - \psi'(0^-)\right]\delta(x) + \left[h_+^{-1}\psi'(0^+) - h_-^{-1}\psi'(0^-)\right]\delta'(x). \end{aligned} \tag{28}$$

The ability to provide explicit forms for the interaction, such as (23), (24), and (28), in terms of well defined distributions (consistent with the equation of motion), is the main difference between the results obtained from the distributional and the SAE approaches. While both methods coincide in the characterization of all point interactions by a four parameter family, only the distributional approach provides such an explicit visualization of the interaction as a distribution concentrated at the origin, a feature which allows one to investigate the properties of the interaction under symmetry transformations in a straightforward way. This will be used in the next section to properly define the notion of a point interaction that is odd under parity – a non-trivial task in the context of SAE (see, for instance, [40] for a broad investigation of the symmetries of the Schrödinger operators with point interactions).

2.1. Symmetry Under Parity Transformations

Let us investigate the behavior of the Schrödinger Equation (2) under parity transformations, in order to select subfamilies of interactions according to their symmetry properties. From (5) and (15) the interaction distribution can be put in the form

$$\begin{aligned} s[\psi](x) &= (\ \delta'(x),\ \ \delta(x)\ )\ \Omega \\ &= (\ \delta'(x),\ \ \delta(x)\ )\ [A_1 - A_2\ U]\ V_1[\psi], \end{aligned} \quad (29)$$

with the matrices A_i's given by (14). Defining $\phi(-x) \equiv \psi(x)$ and noticing that $\psi(0^{\pm}) = \phi(0^{\mp})$ and $\psi'(0^{\pm}) = -\phi'(0^{\mp})$, the Schrödinger Equation (2) can be rewritten in terms of ϕ as

$$\phi''(-x) + k^2\phi(-x) = (\ \delta'(x),\ \ \delta(x)\ )\ [A_1 - A_2\ U]\ \sigma_1\ V_1[\phi], \quad (30)$$

where σ_i ($i = 1, 2, 3$) indicates a Pauli matrix and $V_1[\phi]$ is obtained from (11) by making the substitution $\psi \to \phi$. Performing a space reflection, $x \to -x$, and taking into account that $\delta'(-x) = -\delta'(x)$ and $\delta(-x) = \delta(x)$ the above equation becomes

$$\begin{aligned} \phi''(x) + k^2\phi(x) &= (\ \delta'(x),\ \ \delta(x)\ )\ [-\sigma_3\,(A_1 - A_2\ U)\,\sigma_1]\ V_1[\phi] \\ &\equiv \tilde{s}[\phi](x). \end{aligned} \quad (31)$$

Since $-\sigma_3\, A_j\sigma_1 = A_j$, ($j = 1, 2$), and

$$\tilde{U} = \sigma_1 U \sigma_1 = e^{i\theta}\begin{pmatrix} z^* & -w^* \\ w & z \end{pmatrix}, \quad (32)$$

we have that

$$-\sigma_3\,(A_1 - A_2\ U)\,\sigma_1 = A_1 - A_2\,\tilde{U} \quad (33)$$

Therefore, ϕ solves a Schrödinger equation with an interaction term s˜s˜[ϕ](x) which conserves probability across the origin and it is completely determined by the unitary matrix $\tilde{U}$.

For a regular potential, the interaction is said to be *even* under parity if $\phi(x) = \psi(-x)$ is a solution of the*same* Schrödinger equation solved by $\psi(x)$. Extending this concept to point interactions, we say that an interaction is *even* if s˜s˜[ϕ](x) = s[ϕ](x), which is equivalent to the condition $\tilde{U} = U$. From (32) we see that this condition is satisfied if, and only if, $z = z^*$ and $w = -w^*$. Thus, for $w \neq 0$ Equations (20, 21) imply that the interaction is even if a = d and $\varphi = 0$, whereas for $w = 0$ we must have $h_+ = -h_- < \infty$ or $h_+ = h_- = \infty$.

As an example of an even interaction we mention the well-known δ-interaction, given by $\varphi = 0$, $a = d = 1$, $b = 0$ and c an arbitrary real number. Another example of an even interaction is given by the choice $\varphi = 0$, $a = d = 1$, $c = 0$ and b an arbitrary real number, which corresponds to the interaction term $s[\psi](x) = b\psi'(0^-) = b\psi'(0^+) \equiv b\psi'(0)$, the so-called δ'-interaction – which, as it is well-known, does not have the "correct" (odd) symmetry under parity, as one would expect from an interaction associated to the "potential" δ' (see, for instance, [6, 20, 21]).

Given the "unexpected" behavior of the δ'-interaction under parity transformations, let us investigate whether the four parameter family of general point interactions includes or not any interaction which behaves under parity in the same way that we would expect of an interaction associated with an odd potential. To this end, we recall that for a regular odd potential $V(x)$ [i.e., $V(-x) = -V(x)$] it follows that $\phi(x) = \psi(-x)$ satisfies a Schrödinger equation for an interaction term with the *opposite* sign with respect to the interaction term in the equation satisfied by $\psi(x)$. This is generalized to point interactions by saying that an interaction is *odd* under parity if it satisfies the condition $\tilde{s}\tilde{s}[\phi](x) = -s[\phi](x)$. From (29), (31), and (33) this condition is satisfied if, and only if,

$$A_1 - A_2\,\tilde{U} = -\left[A_1 - A_2\,U\right]. \tag{34}$$

Condition (34) implies that $\Re(z) = 0, \theta = \frac{\pi}{2}$ and $\Im(w) = -1$; these, together with the unitary condition (16), give $\Im(z) = \Re(w) = 0$. Notice that this case ($w = -i \neq 0$) is characterized by a matrix Λ which, according to (20), is the identity matrix. This corresponds to an interaction term which is identically zero and proves that *there is no point interaction in non-relativistic one-dimensional quantum mechanics that behaves properly as an odd interaction*. Therefore, point interactions are either even or have no defined parity under space reflections.

A similar result about the nonexistence of a "genuine" δ'-interaction was obtained in [20] through a*particular* regularization procedure. We stress that our result is mathematically rigorous and *independent of regularization*, thus helping to clarify and to put into perspective any result obtained from a regularization procedure. In fact, suppose that a regularized product $\psi(x)V_\epsilon(x)$ is used as the interaction term into the Schrödinger equation, where V_ϵ is a sequence of regular distributions converging to a point potential $V(x)$ in the zero-range limit $\epsilon \to 0$ (here ϵ may indicate multiple indices). If in this limit the regular products $\psi(x)V_\epsilon(x)$ converge in the sense of distributions, our approach implies that the limiting distribution must be a member (or a subfamily) of the four-parameter family of point interactions described above, which does not contain any genuinely odd interaction. Thus, we conclude that there is no regularization procedure which is able to produce an odd point interaction: even if all the regular products $\psi(x)V_\epsilon(x)$ of the sequence are odd under parity, in the zero-range limit the interaction term will lose this property. As an illustration of this fact, we recall that in [16] Šeba demonstrated that if one considers the $\delta'(x)$ potential as the zero-range limit of the particular sequence of (singular) distributions $V_\epsilon(x) = [\delta(x + \epsilon) - \delta(x - \epsilon)]/2\epsilon$ (for which the product $\psi(x)V_\epsilon(x)$ is well defined for all $\epsilon > 0$), the interaction term obtained in the zero range limit corresponds to an impenetrable barrier, with reflection and transmission coefficients$R = -1$ and $T = 0$ (see also [6])[1]. Notice that for $\epsilon > 0$ each regularized product $\psi(x)V_\epsilon(x)$ behaves as an *odd*interaction, but in the limit $\epsilon \to 0$ these products converge to an interaction of the form (28) with $h_+ = h_- = \infty$, which corresponds to an impenetrable δ-interaction – an *even* interaction under parity transformations.

3. DIRAC'S EQUATION WITH POINT INTERACTIONS

In this section we generalize the approach introduced in the previous section to find the most general family of point interactions allowed in one-dimensional relativistic quantum mechanics, in the context of Dirac's equation. Let us first recall that, in one dimension, the time-independent Dirac equation for a particle of mass m interacting with a *regular* potential $V(x)$ is (throughout this section we adopt natural units, $\hbar = c = 1$)

$$\left(i\alpha_x \frac{d}{dx} - \beta m + E\right)\psi(x) = \psi(x)V(x), \tag{35}$$

where α_x and β are the Dirac matrices satisfying $\{\alpha_x, \beta\} = 0$ and which can be chosen as $\beta = \sigma_3$ and $\alpha_x = \sigma_1$, with σ_i indicating the Pauli's matrices. The Dirac spinor $\psi(x)$ can be written in terms of its two components as $\psi(x) = (u(x), v(x))^T$, where T stands for the transpose.

As it is well known, in the relativistic case even for the simplest point interaction (i.e., the Dirac delta potential) the interaction term cannot be naively defined as the product $\psi(x)V(x)$: defining the interaction as $\psi(0)\delta(x)$ is inconsistent with the fact that Dirac's equation implies that $\psi(x)$ must be discontinuous at the origin, as can be seen by equating the singular order in both sides of the resulting equation. Thus, in the same spirit of the previous section, we rewrite Dirac's Equation (35) with the interaction term substituted by a well-defined interaction distribution,

$$\left(i\alpha_x \frac{d}{dx} - \beta m + E\right)\psi(x) = S[\psi](x)\,, \tag{36}$$

where $S[\psi](x)$ is a two-component spinor whose components are well defined distributions. Similarly, we demand that requirement (*i*) and (*ii*) of the previous section be satisfied. Together with Theorem 1 these imply that

$$S[\psi](x) = \sum_{n=0}^{r_s} \Omega_n[\psi]\,\delta^{(n)}(x)\,, \tag{37}$$

where r_s is the singular order of the interaction term and $\Omega_n[\psi] = (\varphi_n^0[\psi],\ \varphi_n^1[\psi])^T$ are spinors whose (yet to be determined) components are linear functionals of the boundary values $\psi(0^{\pm})$, which completely characterize the interaction. However, the most singular point interaction allowed in the one dimensional Dirac's equation (which still gives the possibility to build normalizable wave functions as superpositions of the stationary solutions) is described by a distribution with singular order zero, corresponding to a δ interaction. Thus, from now on we restrict ourselves to consider Equation (37) with $r_s = 0$, which is then reduced to

$$S[\psi](x) = \Omega_0[\psi]\,\delta(x)\;; \qquad \Omega_0[\psi] = \begin{pmatrix} \varphi^0 \\ \varphi^1 \end{pmatrix}; \tag{38}$$

where we dropped the subscript "0" in the components $\varphi^{0(1)}$, to simplify the notation.

Substituting the interaction (38) into Dirac's equation results in b.c. for the Dirac spinor given by

$$\psi(0^+) - \psi(0^-) = -i\alpha_x\Omega_0[\psi] \tag{39}$$

as can be seen by taking an indefinite integral of the Dirac equation and using the fact that the primitive $\psi^{(-1)}$ is continuous (since $r_s = 0$, this primitive has singular order −2).

The equation resulting from the substitution of (38) into (36) can be rewritten as two coupled first order equations for $u(x)$ and $v(x)$, which in turn can be reduced to

$$v(x) = \frac{1}{(E+m)}\left[\varphi^1\delta(x) - iu'(x)\right], \tag{40}$$

and a second order equation for $u(x)$:

$$u''(x) + k_r^2 u(x) = (E+m)\varphi^0\,\delta(x) - i\varphi^1\,\delta'(x), \tag{41}$$

with $k_r \equiv \sqrt{E^2 - m^2}$. Equation (41) is a Schrödinger-like equation for $u(x)$ with a singular interaction of order +1.

The solution of (40, 41) follows directly from the results in section 2. However, before proceeding to obtain the explicit form of the interaction term, let us first consider the requirement (*iii*) of probability current conservation. In one dimension the relativistic current density is given by $j^1(x) = \psi^\dagger(x)\alpha_x\psi(x) = u^*(x)v(x) + u(x)v^*(x)$; from Equation (40) it follows that $v(0^\pm) = -\frac{i}{(E+m)}u'(0^\pm)$ and therefore

$$j^1(0^\pm) = \frac{1}{i(E+m)}\left\{u^*(0^\pm)u'(0^\pm) - u^{*\prime}(0^\pm)u(0^\pm)\right\}. \tag{42}$$

Hence, conservation of the "non-relativistic" probability current associated with $u(x)$ implies the conservation of the relativistic current across the singular point.

The formal similarity between (41) and the Schrödinger equation with the interaction term (5) allows us to find the b.c. for u and u' at the origin by identifying $(E + m)\varphi^0 = \alpha_0$ and $i\varphi^1 = \alpha_1$, so that all the results from the previous section are also valid for the u-component of the Dirac spinor. Then, by taking into account (40) we obtain the b.c. for v and for the Dirac spinor ψ. In fact, it follows from (18) that the vector $\Upsilon(x) \equiv (u(x), u'(x))^T$ satisfies the b.c. $R_+\Upsilon(0^+) = R_-\Upsilon(0^-)$, with $R_\pm$ given in section 2. Noticing that $\Upsilon(0^\pm) = G\psi(0^\pm)$, with the G matrix defined as

$$G = \begin{pmatrix} 1 & 0 \\ 0 & i(E+m) \end{pmatrix},$$

from the results for the non-relativistic theory it follows directly that:

(a) If $\det R_\pm \neq 0$ the relativistic b.c. can be written as

$$\psi(0^+) = \Lambda_r\,\psi(0^-), \tag{43}$$

where $\Lambda_r = G^{-1}\Lambda G$ can be written as (the subscript "r" is to indicate relativistic quantities)

$$\Lambda_r = e^{i\varphi_r}\begin{pmatrix} a_r & ib_r \\ -ic_r & d_r \end{pmatrix}, \quad a_r d_r - b_r c_r = 1, \tag{44}$$

with $\varphi_r \in [0, \pi)$ and $a_r, b_r, c_r, d_r \in \mathbb{R}$ are (dimensionless) constants.

(b) If det $R_\pm = 0$, the relativistic b.c. can be written as

$$v\left(0^{\pm}\right) = ih_r^{\pm} u\left(0^{\pm}\right), \quad h_r^{\pm} \in \mathbb{R} \cup \{\infty\}. \tag{45}$$

Therefore, we conclude that, similarly to the non-relativistic case, there are two possible situations in one-dimensional relativistic quantum mechanics: for *non-separated* solutions [case (*a*) above] there is a four-parameter family of interactions that includes all the possible point interactions, and for *separated*solutions [case (*b*) above; corresponding to set $w = 0$ in (16)] the number of free parameters is reduced to two. Again, this result agrees with that obtained by SAE [32] – for works dealing with particular one-parameter subfamilies of these interactions see [27–30].

The explicit form of the relativistic interaction term for the case (*a*) above can be found from (38), (39), and (43) as

$$S[\psi](x) = i\alpha_x \left(\Lambda_r - 1\right) \psi(0^-)\, \delta(x), \tag{46}$$

whereas for the case (*b*) we use (38), (39), and (45) to obtain

$$S[\psi](x) = i\alpha_x \left[\begin{pmatrix} 1 & 0 \\ ih_r^+ & 0 \end{pmatrix} \psi(0^+) - \begin{pmatrix} 1 & 0 \\ ih_r^- & 0 \end{pmatrix} \psi(0^-) \right] \delta(x). \tag{47}$$

It is convenient to express (41) explicitly in terms of the parameters in (44) or (45). For case (*a*), with the interaction term given by (46), the *u*-component of the Dirac spinor satisfies [also see Equation (38)]

$$\begin{aligned} u''(x) + k_r\, u(x) &= \left[(E+m)c_r\, e^{i\varphi_r} u(0^-) + \left(d_r e^{i\varphi_r} - 1\right) u'(0^-)\right] \delta(x) \\ &+ \left[\left(a_r e^{i\varphi_r} - 1\right) u(0^-) + \frac{b_r}{E+m} e^{i\varphi_r} u'(0^-)\right] \delta'(x). \end{aligned} \tag{48}$$

Similarly, when the interaction term is given by (47) [case (b)] the equation for the *u*-component reads

$$u''(x) + k_r^2 u(x) = (E+m)\left[-h_r^+ u(0^+) + h_r^- u(0^-)\right]\delta(x) + \left[u(0^+) - u(0^-)\right]\delta'(x). \tag{49}$$

Equations (48, 49) are suitable to analyze the non-relativistic limit (characterized, as usual, by $E \to m$, $k_r \to k$) of Dirac's equation. Following a procedure analogous to the one presented in [43], it can be shown that in the non-relativistic limit $v(x)$ is the "small" component of the Dirac spinor, whereas the above Schrödinger-like equations satisfied by the "large" component $u(x)$ correspond, in that limit, to Equations (23, 27), respectively. Comparing the above equations with their equivalent in the non-relativistic case [see Equations (23, 28), respectively], we can establish a one-to-one correspondence between the parameters of the relativistic and non-relativistic families of point interactions:

$$\varphi = \varphi_r, \quad a = a_r, \quad b = \frac{b_r}{2m}, \quad c = 2m\, c_r, \quad d = d_r, \tag{50}$$

$$h_\pm = -2m\, h_r^\pm, \tag{51}$$

which was also reported in the context of SAE [32]. For instance, from these relations it is straightforward to see that the relativistic interaction that corresponds in the non-relativistic limit to a δ-interaction with strength γ is given by $\varphi_r = b_r = 0$, $a_r = d_r = 1$, $c_r = \gamma$, which corresponds to the following interaction term in the Dirac equation

$$S[\psi](x) = \frac{1}{2}\gamma(\mathbf{1} + \beta)\psi(0^-)\delta(x), \tag{52}$$

where **1** is the 2× 2 identity matrix. This interaction is an equal mix of electrostatic and scalar potentials, which is known to be a confining interaction [29] (i.e., the transmission amplitude approaches zero in the limit $\gamma \to \infty$). On the other hand, the relativistic interaction given by $\varphi_r = 0$, $a_r = d_r = 1$, $b_r = -\gamma$ and $c_r = 0$,

$$S[\psi](x) = \frac{1}{2}\gamma(\mathbf{1} - \beta)\psi(0^-)\delta(x), \tag{53}$$

is the inverted mix of electrostatic and scalar potentials (also confining [29]) and corresponds, in the non-relativistic limit, to the interaction given by $\varphi = c = 0$, $a = d = 1$ and b=−γ2mb=−γ2m, which is the so-called δ'-interaction (with strength $-\gamma/4m^2$). Other well-known one-parameter subfamilies of relativistic interactions are the pure scalar ($\varphi_r = 0$, $b_r = c_r = \gamma$, $a_r = d_r = 1$) and pure electrostatic ($\varphi_r = 0$, $-b_r = c_r = \gamma$, $a_r = d_r = 1$), which do not correspond in the non-relativistic limit either to a δ or to the δ' interaction, as already observed by Benvegnù and Dąbrowski [32]. These are, of course, just some examples of particular one-parameter subfamilies often studied in the framework of relativistic quantum mechanics in one dimension; all of them are included in the four-parameter family of interactions considered above.

Finally, the results of subsection 2.1, concerning the characterization of non-relativistic point interactions with respect to their behavior under space reversal (parity transformations), can be straightforwardly extended to relativistic interactions. Accordingly, if $\psi(x)$ solves a Dirac equation with the interaction $S[\psi](x)$, the interaction is said to be *even* under space reversal if $\tilde{\psi}\tilde{\psi}(x) \equiv \beta\psi(-x)$ solves the same equation, and it is said to be *odd* if $\tilde{\psi}\tilde{\psi}(x)$ solves the Dirac equation for an interaction term with *reversed* sign, $-S[\tilde{\psi}\tilde{\psi}](x)$. The conditions that the relativistic parameters must satisfy to characterize an even interaction are $\varphi_r = 0$, $a_r = d_r$ for the non separated solutions, and $h^+_r = -h^-_r < \infty$ or $h^+_r = h^-_r = \infty$ for the separated solutions. These are exactly the conditions that we would obtain from the correspondence relationships between the non-relativistic and relativistic parameters, Equations (50, 51). In a similar way, we conclude that there is no point interaction which is odd under space reversal in one dimensional relativistic quantum mechanics.

4. CONCLUDING REMARKS

In this work we considered a distributional approach to treat singular point potentials in one-dimensional quantum mechanics. The method relies on the fact that any singular distribution concentrated at a point must be a linear combination of the δ-distribution and its derivatives (see Theorem 1). This, together with requirement (*ii*) (section 2), allows us to substitute the ill-defined product $V(x)\psi(x)$ in the Schrödinger and

Dirac equations by a well-defined interaction distribution with a given singular order. Then, the interaction distribution is completely determined by imposing the physical requirement that the probability current is conserved across the singularity—notice that even though the probability current is, in general, ill-defined *at* the singularity, it is well-defined on both sides of the singularity, so that in all the steps we only deal with well defined distributions. Implicit in the method is a *locality condition*, i.e., the interaction term depends both on singular potential and the wave function at the vicinities of the singular point.

The application of the distributional approach to both non-relativistic and relativistic quantum mechanics leads to a four-parameter family of point interactions, a result in agreement with that obtained by the method of self-adjoint extensions. It is important to notice that, while our results coincide with those of SAE in what regards the characterization of the family of all point interactions for Schrödinger's and Dirac's theories, there is an important conceptual difference between the two methods. While in SAE the interaction is defined only in terms of the boundary conditions around the singular point, in the present distributional approach the interaction is *explicitly constructed* from simple mathematical and physical requirements as an *interaction distribution* concentrated at the origin, which implies the boundary conditions satisfied by the wave function.

The fact that the interaction term is explicitly provided [following from the requirements (*i*)-(*iii*) in section 2] makes it particularly simple to analyze the symmetries of the interaction, as the investigation of parity transformations in sections 2, 3 shows. As one of the main results of the distributional approach such an analysis shows, in a simple way, why there is no *odd* point interaction in one dimension. Indeed, an immediate general consequence of the distributional method is that any regularization using δ'-converging sequences of functions for the potential must converge, *in the sense of distributions*, to an interaction distribution *which cannot be odd* under parity. We conjecture that a failure to satisfy this criterion is at the origin of the sometimes contradictory results appearing in the literature of regularization based treatments of zero-range interactions. Notice that, in general, these regularization methods assume from the start odd sequences of short-range potentials with the expectation of obtaining an "odd δ'-interaction" in the zero-range limit (even though, of course, this is no guarantee of convergence to an odd distribution, as clearly demonstrated by Šeba [16]).

Finally, we note that the distributional approach to point interactions developed here offers a mathematically rigorous and less abstract alternative to the (also rigorous) SAE method. In addition, the method is expected to generalize in a straightforward way to higher dimensions as well as to more general interactions in one or more dimensions which still present open questions, such as the Coulomb potential [44]. Preliminary work applying the distributional approach to the one-dimensional Coulomb interaction suggests that the fact that one obtains an explicit form for the interaction distribution, together with symmetry considerations, may help to clarify some ambiguities concerning a proper characterization of this interaction.

CONFLICT OF INTEREST STATEMENT

The authors declare that the research was conducted in the absence of any commercial or financial relationships that could be construed as a potential conflict of interest.

ACKNOWLEDGMENTS

Wagner Monteiro thanks Fundaç~ao Araucária for financial support. Luiz A. Manzoni thanks the Office of Undergraduate Research, Concordia College, for partial financial support. José T. Lunardi thanks F. M. de Andrade for stimulating discussions at the beginning of this work.

FOOTNOTES

1. ^Šeba [16] has also shown that the same result holds if one uses as $V_\epsilon(x)$ a sequence of infinitely smooth functions converging to $\delta'(x)$ in the sense of distributions.

REFERENCES

1. Albeverio S, Gesztesy F, Høegh-Krohn R, Holden H. *Solvable Models in Quantum Mechanics*. 2nd ed. Providence, RI: AMS Chelsea Publishing (2004).

2. Coutinho FAB, Nogami Y, Tomio L. Many-body system with a four-parameter family of point interactions in one dimension. *J Phys A Math Gen*. (1999) **32**:4931–42.

3. Schmidt AGM, Cheng BK, da Luz MGE. Green functions for generalized point interactions in one dimension: a scattering approach.*Phys Rev A* (2002) **66**:062712.

4. Gadella M, Negro J, Nieto LM. Bound states and scattering coefficients of the $-a\delta(x)$ $+b\delta'(x)$ potential. *Phys Lett A* (2009) **373**:1310–13.

5. Arnbak H, Christiansen PL, Gaididei YB. Non-relativistic and relativistic scattering by short-range potentials. *Philos Trans R Soc A*(2011) **369**:1228–44. 6. Coutinho FAB, Nogami Y, Perez JF. Generalized point interactions in one-dimensional quantum mechanics.*J Phys A Math Gen*. (1997)**30**:3937–45.

6. Zolotaryuk AV. Boundary conditions for the states with resonant tunnelling across the δ'-potential. *Phys Lett A* (2010) **374**:1636–41.

7. Lieb EH, Liniger W. Exact analysis of an interacting Bose Gas. I. The general solution and the ground state. *Phys Rev*. (1963)**130**:1605–16.

8. Lieb EH. Exact analysis of an interacting Bose gas. II. The excitation spectrum. *Phys Rev.* (1963) **130**:1616–24.

9. Exner P. Lattice Kronig-Penney models. *Phys Rev Lett.* (1995) **74**:3503–06.

10. Kundu A. Exact solution of double δ function Bose gas through an interacting anyon gas. *Phys Rev Lett.* (1999) **83**:1275–78.

11. Comtet A, Texier C, Tourigny Y. Products of random matrices and generalised quantum point scatterers. *J Stat Phys.* (2010)**140**:427–66.

12. Jackiw R. Delta-function potentials in two- and three-dimensional quantum mechanics. In: Ali A, Hoodbhoy P, editors. *M. A. B. Bég Memorial Volume.* Singapore: World Scientific (1991). p. 35–53.

13. Manuel C, Tarrach R. Perturbative renormalization in quantum mechanics. *Phys Lett B* (1994) **328**:113–8.

14. Nyeo SL. Regularization methods for delta-function potential in two-dimensional quantum mechanics. *Am J Phys.* (2000) **68**:571–5.

15. Šeba P. Some remarks on the δ-interaction in one dimension. *Rep Math Phys.* (1986) **24**:111–20.

16. Zhao BH. Comments on the Schrödinger equation with δ′-interaction in one dimension. *J Phys A Math Gen.* (1992) **25**:L617–8.

17. Griffiths DJ. Boundary conditions at the derivative of a delta function. *J Phys A Math Gen.* (1993) **26**:2265–67.

18. Albeverio S, Gesztesy F, Holden H. Comments on a recent note on the Schrödinger equation with a δ′-interaction. *J Phys A Math Gen.* (1993) **26**:3903–4.

19. Román JM, Tarrach R. The regulated four-parameter one-dimensional point interaction. *J Phys A Math Gen.* (1996) **29**:6073–85.

20. Cheon T, Shigehara T. Realizing discontinuous wave functions with renormalized short-range potentials. *Phys Lett A* (1998) **243**:111–6.

21. Griffiths D, Walborn S. Dirac deltas and discontinuous functions. *Am J Phys.* (1999) **67**:446–7.

22. Šeba P. The generalized point interaction in one dimension. *Czech J Phys B* (1986) **36**:667–73

23. Exner P. Contact interactions on graph superlattices. *J Phys A Math Gen.* (1996) **29**:87–102.

24. Carreau M. Four-parameter point-interaction in 1D quantum systems. *J Phys A Math Gen.* (1993) **26**:427–32.

25. Christiansen PL, Arnbak HC, Zolotaryuk AV, Ermakov VN, Gaididei YB. On the existence of resonances in the transmission probability for interactions arising from derivatives of Dirac's delta function. *J Phys A Math Gen.* (2003) **36**:7589–600.

26. Sutherland B, Mattis DC. Ambiguities with the relativistic δ-function potential. *Phys Rev A* (1981) **24**:1194–97.

27. McKellar BHJ, Stephenson GJ. Klein paradox and the Dirac-Kronig-Penney model. *Phys Rev A* (1987) **36**:2566–9.

28. Dominguez-Adame F, Macia E. Bound states and confining properties of relativistic point interaction potentials. *J Phys A Math Gen.* (1989) **22**:L419–23.

29. Roy CL. Boundary conditions across a δ-function potential in the one-dimensional Dirac equation. *Phys Rev A* (1993) **47**:3417–19.

30. Falkensteiner P, Grosse H. Quantization of fermions interacting with point-like external fields. *Lett Math Phys.* (1987) **14**:139–48.

31. Benvegnù S, Dąbrowski L. Relativistic point interaction. *Lett Math Phys.* (1994) **30**:159–67.

32. Epstein H, Glaser V. The role of locality in perturbation theory. *Ann Inst H Poincaré A* (1973) **19**:211–95.

33. Scharf G. *Finite Quantum Electrodynamics: The Causal Approach*. 2nd ed. Berlin: Springer (1995).

34. Manzoni LA, Pimentel BM, Tomazelli JL. Causal theory for the gauged thirring model. *Eur Phys J C* (1999) **8**:353–61.

35. Lunardi JT, Pimentel BM, Valverde JS, Manzoni LA. Duffin-Kemmer-Petiau theory in the causal approach. *Int J Mod Phys A* (2002)**17**:205–27.

36. Lunardi JT, Manzoni LA, Monteiro W. Remarks on point interactions in quantum mechanics. *J Phys Conf Ser*. (2013) **410**:012072.

37. Zemanian AH. *Distribution Theory and Transform Analysis: An Introduction to Generalized Functions, with Applications*. New York, NY: Dover Publications (1987).

38. Gel'fand IM, Shilov GE. *Generalized Functions: Properties and Operations*. Vol. 1. New York, NY: Academic Press (1964).

39. Albeverio S, Dabrowski L, Kurasov P. Symmetries of Schrödinger operator with point interactions. *Lett Math Phys*. (1998) **45**:33–47.

40. Kurasov P. Distribution theory for discontinuous test functions and differential operators with generalized coefficients. *J Math Anal Appl*. (1996) **201**:297–323.

41. Fülöp T, Tsutsui I. A free particle on a circle with point interaction. *Phys Lett A* (2000) **264**:366–74.

42. Bjorken JD, Drell SD. *Relativistic Quantum Mechanics*. New York, NY: McGraw-Hill (1964).

43. de Oliveira CR, Verri AA. Self-adjoint extensions of Coulomb systems in 1, 2 and 3 dimensions. *Ann Phys*. (2009) **324**:251–66.

8

THE THEORETICAL RAMIFICATIONS OF THE COMPUTATIONAL UNIFIED FIELD THEORY

Jonathan Bentwich

Brain Perfection LTD, Israel

1. INTRODUCTION

Four previous articles (Bentwich, 2012: a-d) have postulated the existence of a novel 'Computational Unified Field Theory' (CUFT) which is a candidate 'Theory of Everything' (Brumfiel, 2006; Ellis, 1986; Greene, 2003) – i.e., has the potential of unifying between quantum (Born, 1954) and relativistic models of physical reality (and moreover also possesses the potential of opening new 'vistas' for scientific examination connected with its discovery of a new hypothetical 'Universal Computational Principle which carries out an extremely rapid computation, c^2/h of a series of Universal Simultaneous Computational Frames, 'USCF's, which give rise to all 'apparent' secondary computational 'physical' properties of 'space', 'time', 'energy' and 'mass'); Indeed, the primary focus of the current manuscript is precisely to explore the potential theoretical ramifications of this novel CUFT – based on the recognition that the (singular) Universal Computational Principle ('י') solely produces all (apparent) secondary computational 'physical' properties of 'space, 'time', 'energy' and 'mass', and hence comprises the sole "reality" (which produces all exhaustive hypothetical inductive and deductive phenomenon through a higher-ordered 'a-causal' computational framework; this may subsequently bear significant theoretical ramifications for all (key) 'material-causal' scientific paradigms as well as point at the discovery of a (new) *'Universal Consciousness Principle Computational Program'*, as well as outline potential resolutions of major Physical 'enigma's;

Hence, the current manuscript traces the potential theoretical ramifications of:

a. An 'a-causal' computational framework of the (CUFT's) singular Universal Consciousness Principle's ('י') responsible for the (higher-ordered) computation of all exhaustive hypothetical (e.g., empirically knowable) inductive or deductive 'x-y' pairs series – which leads to the discovery of a-*causal 'Universal Consciousness Principle Computational Program'*.

b. An exploration of the CUFT's Universal Consciousness Principle's ('י') and Duality Principle's (Bentwich, 2003c, 2004, 2006) reformalization of all (apparent inductive or deductive) major SROCS computational paradigms (e.g., including: Darwin's 'Natural Selection Principle' (Darwin, 1859) and associated Genetic Encoding hypothesis, Neuroscience's Psychophysical Problem of human Consciousness and all inductive and deductive Gödel-like SROCS paradigms).

c. Theoretical Ramifications of the Universal Consciousness Principle.

2. A SINGULAR 'A-CAUSAL' UNIVERSAL CONSCIOUSNESS PRINCIPLE COMPUTATION OF ALL INDUCTIVE AND DEDUCTIVE 'X-Y' RELATIONSHIPS

We thus begin with an exploration of three potential theoretical ramifications of the CUFT's description of the operation of the (singular)

Universal Consciousness Principle ('י') which has been shown to compute an extremely rapid series of Universal Simultaneous Computational Frames (USCF's);

The Universal Computational/Consciousness Principle was (previously) shown to encapsulate a singular higher-ordered 'D2' computation of an 'a-causal' computation of the "simultaneous co-occurrences" of all exhaustive hypothetical inductive or deductive (e.g., empirically knowable) 'x-y' pairs series; Therefore, the acceptance of the CUFT's description of the Universal Consciousness Principle necessarily implies that throughout the various (inductive or deductive) disciplines of Science we need to shift from the current basic (Cartesian) "material-causal" scientific theoretical towards a singular (higher-ordered 'D2') *'Universal Consciousness Principle's a-causal computation'*:

This means that the current (Cartesian) 'material-causal' scientific framework assumes that any given 'y' element (or value) can be explained as a result of its (direct or indirect) 'causal' interaction/s with another (exhaustive hypothetical inductive or deductive) series of 'x' factor/s – which determines whether that 'y' element (or value) "exists" or "doesn't exist", thereby comprising a 'Self-Referential Ontological Computational System' (SROCS) (Bentwich: 2012a-d):

SROCS: PR{x,y}→ ['y' or 'not y']/di1...din.

But, since it was previously shown that such SROCS computational structure inevitably leads to both 'logical inconsistency' and 'computational indeterminacy' that were shown to be contradicted by robust empirical findings indicating the capacity of the major scientific SROCS paradigms to be capable of determining the "existence" or "non-existence" of the particular 'y' element, see Bentwich 2012b) – then the CUFT's 'Duality Principle' asserted the existence of the singular 'Universal Consciousness Principle' ('י') which is capable of computing the "simultaneous co-occurrences" of any particular (exhaustive hypothetical) 'x-y' pairs series which are embedded within the Universal Computational/Consciousness Principle's rapid series of USCF's.

What this means is that both specifically for each of the (previously identified) key scientific SROCS paradigms as well as more generally for any hypothetical ('empirically knowable') inductive or deductive ('x-y') phenomenon, we must reformulate our scientific understanding in such a way which will allow us to present any such 'x-y' relationship/s as being computed by the singular Universal Consciousness Principle (e.g., as the computation of an exhaustive-hypothetical "co-occurring" 'x-y' pairs' series); In that respect, this (novel) 'Universal Consciousness Principle's' scientific framework shifts Science from its current basic (Cartesian) assumption wherein all natural phenomena can be described as 'material-causal' ('x→y') relationships (e.g., comprising the apparent SROCS computational structure contradicted by the computational Duality Principle) – to an 'a-causal' singular Universal Consciousness Principle which computes the simultaneous "co-occurrences" of any inductive or deductive 'x-y' pairs series comprising the various 'pixels' of the USCF's frames (e.g., produced by this Universal Consciousness Principle).

Finally, it should be noted that a key principle underlying this shift from the current 'material-causal' (Cartesian) scientific framework towards the CUFT's (proven) higher-ordered singular Universal Consciousness Principle's ('י') 'a-causal' theoretical framework is the acceptance of the impossibility of the existence of any such 'material-causal' ('x-y') relationship/s – i.e., due to the impossibility of any 'physical' entity, attribute (or property) being transferred across any (two subsequent) 'USCF's frames: Thus, apart from the (previously shown) conceptual computational proof of the 'Duality Principle' wherein due to the inevitable 'logical inconsistency' and 'computational indeterminacy' arising from the SROCS computational structure (which is contradicted by empirical evidence indicating the capacity of these key scientific SROCS paradigms to compute the "existence" or "non-existence" of any particular 'y' element or value) – which points at the existence of the higher-ordered (singular) 'Universal Computational/Consciousness Principle that computes the "simultaneous co-occurrences" of any (exhaustive-hypothetical) 'x-y' pairs' series; it is suggested that the inclusion of this computational Duality Principle as one of the (seven) theoretical postulates of the CUFT (e.g., specifically alongside the CUFT's

'Computational Invariance' and 'Universal Consciousness' postulates) unequivocally asserts that there cannot (in principle) exist any 'material-causal' effect/s (or relationship/s) being transferred across any (two subsequent) USCF's frames! This is because the CUFT's very definition of all four 'physical' properties of 'space', 'time', 'energy' and 'mass' – as secondary computational by-products of the (singular) Universal Computational Consciousness' computation of (an extremely rapid series of) 'Universal Simultaneous Computational Frames' (USCF's); and moreover the CUFT's 'Computational Invariance' postulate indication that due to the 'computational variance' of these four (secondary computational) 'physical' properties (e.g., as existing only "during" the appearance of the USCF frames but 'non-existence' "in-between" any two such subsequent frames, see Bentwich, 2012:c-d) as opposed to the 'computational invariance' of the 'Universal Consciousness Principle' ('י'), we need to regard only this singular (computationally invariant) 'Universal Consciousness Principle' as "real" whereas all four (secondary computationally variant) 'physical' properties must be regarded as merely 'phenomenal' (i.e., as being comprised in reality only from the singular Universal Consciousness Principle); Therefore, the CUFT's 'Universal Consciousness Principle' advocated that none of these four (secondary computationally variant) 'physical' properties (e.g., of 'space', 'time', 'energy' or 'mass') "really" exists – but rather that there is only this one singular Universal Consciousness Principle which exists (solely) "in-between" any (two subsequent) USCF's frames and also solely produces each of these USCF's derived four 'phenomenal physical' properties; Hence, it was evinced (by the CUFT's Universal Consciousness Principle) that there cannot be any 'transference' of any hypothetical 'material' or 'physical' entity, effect, or property across any (two subsequent) USCF's frames! We therefore reach the inevitable theoretical conclusion that the current scientific (Cartesian) "material-causality' basic assumption underlying all key scientific SROCS paradigms as well as all (empirically knowable) 'Gödel-like' (inductive or deductive) SROCS 'x-y' relationships, wherein there exists a 'material-causal' effect/s (or relationship/s) between any given 'x' element and any (exhaustive hypothetical) 'y' series which determines the "existence" or "non-existence" of that (particular) 'y' element (or value) – is untenable! Instead, we must accept the CUFT's assertion that there can only exists one singular 'Universal Consciousness Principle' ('י') which both (solely) produces- all (apparent) secondary computational 'physical' properties (of 'space', 'time', 'energy' and 'mass'), as well as computes the "simultaneous co-occurrences" of any (particular) exhaustive-hypothetical inductive or deductive 'x-y' pairs series (e.g., comprising the exhaustive USCF's frames).

3. THE "UNIVERSAL CONSCIOUSNESS PRINCIPLE'S COMPUTATIONAL PROGRAM"

Therefore, it follows that based on the recognition of the singularity of the Universal Consciousness Principle's 'a-casual' computation of the "simultaneous co-occurrences" of all (inductive or deductive) 'x-y' pairs' series (as comprising the exhaustive USCF's frames) – we need to be able to reformulate all of the previously mentioned key scientific SROCS paradigms (Bentwich, 2012b-d), including: Darwin's 'Natural Selection Principle' and associated 'Genetic Encoding' hypothesis, Neuroscience's Psychophysical Problem of human Consciousness, as well as all (exhaustive hypothetical) 'Gödel-like' (apparent) inductive or deductive SROCS computational paradigms based on this singular (higher-ordered) Universal Consciousness Principle's ('י') 'a-causal' USCF's computation;

Hence, what follows is a description of the principle theoretical ramifications of reformulating each of these key scientific (apparent) SROCS computational paradigms, as well as a more generalized description of a tentative 'Universal Consciousness Principle Program' (e.g., which may offer a successful alternative for 'Hilbert's Mathematical Program' to base all of our human scientific knowledge upon the foundations of the operation of the singular Universal Consciousness Principle). First, it may be worthwhile to rearticulate the reformalization of each of these key scientific (apparent) SROCS paradigms in terms of the operation of the singular Universal Consciousness Principle (as previously outlined: Bentwich, 2012b):

S.:**D2**: [{E{1...n}, o}st1; {E{1...n}, o}st2... {E{1...n}, o}stn].

G.F – P.S.:**D2**: [{G{1...n}, 'ph*i (o)*' }st1; {G{1...n}, 'ph*j (o)*' }sti;...{G{1...n}, 'ph*n(o)*' }stn].

G.E. – P.S.:**D2**: [{G*e*{1...n}, p*i-synth* (o-phi)}st1; G*e*{1...n}, p*j-synth* (o-phi)}sti... ; G*e*{1...n}, p*n-synth* (o-phi)}stn]

Psychophysical:**D2**: [{N(1…n)$_{st-i}$, Cs-pp $_{st-i}$}; … {N(1…n)$_{st-i+n}$, Cs-pp $_{st-i+n}$ }]

Functional: **D2**: [{Cs(pp)f$_{i}$, Na(spp)fi}$_{st-i}$; … {Cs(pp)f$_{(i+n)}$, Na(spp)f$_{(i+n)}$} $_{st(i+n)}$]

Phen.: **D2**: [{Cs(pp- fi)-Ph$_{i}$, Na(spp-fi)-Ph$_{i}$} $_{st-i}$; …{Cs(pp- fi)-Ph$_{(i+n)}$, Na(spp-fi)-Ph} $_{st-(i+n)}$]

Self: **D2**: [{Cs(pp- fi)Ph-Si, Na(pp- fi)Ph-S$_{i}$} $_{st-i}$; …{Cs(pp- fi)Ph-S(i+n), Na(pp- fi)Ph-S(i+n)} $_{st-(i+n)}$]

GIT:**D2**: ([{S{1...n}, t}*i* … {S{1...n}, t}*z*], or [{x{1...n}*i*, y*i*} … {x{1...n}*z*, y*z*}])

Indeed, what may be seen from this singular description of all of these key scientific SROCS paradigms, is that it recognize the fact that all of these major (apparent) SROCS paradigms are computed simultaneously as different "co-occurring" 'x-y' pairs embedded within the same (single or multiple) USCF frame that is produced by the singular Universal Consciousness Principle ('ב'); What this means is that the recognition of the singularity of this Universal Consciousness Principle as the sole "reality" which computes the "simultaneous co-occurrences" of all of these (particular) exhaustive hypothetical 'x-y' pairs series, and which also exists (solely) "in-between" any two such USCF's – forces us to transcend the 'narrow constraints' of the (current) Cartesian 'material-causal' theoretical framework (e.g., which assumes that any given 'y' entity (or phenomenon) is "caused" by its (direct or indirect) physical interaction/s with (an exhaustive hypothetical 'x' series); Instead, this singular Universal Consciousness Principle 'a-causal' computation asserts that it is the same singular Universal Consciousness Principle which computes- produces- retains- and evolves- all of these particular scientific (apparent) SROCS 'x-y' pairs series across a series of USCF's...

In other words, instead of the existence of any "real" material-causal relationship between any of these (particular SROCS) 'x→y' entities (e.g., Darwin's Natural Selection Principle's assumed 'material-causal' relationship between an organism's Environmental Factors, 'x', and own traits or behavior 'y'; or between any exhaustive hypothetical Genetic Factors and any given phenotypic behavior; or between Neuroscience's Psychophysical Problem of Human Consciousness' psychophysical stimulation, 'x', and Neural Activation, 'y'; or in fact between any hypothetical inductive or deductive Gödel-like SROCS 'x-y' factors); the CUFT's Universal Consciousness Principle offers an alternative singular (higher-ordered) computational mechanism which computes the "simultaneous co-occurrences" of any of these (exhaustive hypothetical) 'x-y' pairs' series – which are all produced- and embedded- within the Universal Consciousness Principle's computed USCF's frames... Indeed, the shift from the current 'material-causal' (Cartesian) scientific framework towards the Universal Consciousness Principle's singular computation of the "simultaneous co-occurrences" of all exhaustive hypothetical (inductive or deductive) 'x-y' pairs' series may lead the way for reformulating all of these key scientific SROCS paradigms (as well as any other hypothetical inductive or deductive 'x-y' series) within a basic "Universal Consciousness Principle Computational Program";

Essentially, such a *'Universal Consciousness Principle's Computational Program'* is based upon the foundations of the CUFT's (abovementioned) three postulates of the 'Duality Principle', the 'Computational Invariance' principle and the 'Universal Consciousness Principle' – all pointing at the fact that all empirically computable (inductive or deductive) 'x-y' relationships must necessarily be based upon the singular (conceptually higher-ordered) Universal Consciousness Principle which is solely responsible for the computation of the "simultaneous co-occurrences" of all such (exhaustive hypothetical) inductive or deductive 'x-y' pairs series comprising the totality of the USCF's (single or multiple) frames.... Moreover, this singular Universal Consciousness Principle ('י') was also shown to exist independently of any (secondary computational) 'physical properties' (e.g., of 'space', 'time', 'energy' and 'mass') and therefore constitute the only "reality" that exists invariantly (i.e., both as giving rise to the four 'phenomenal' physical properties and as existing solely "in-between" any two such subsequent USCF's frames).

In order to appreciate the full (potential) theoretical significance of such a 'Universal Consciousness Principle Computational Program' it may be worthwhile to reexamine Hillbert's famous 'Mathematical Program' to base Mathematics upon the foundations of Logic (e.g., and by extension also all of Science upon the foundations of Mathematics and Logic), and more specifically, to revisit 'Gödel's Incompleteness Theorem' (GIT) which delivered a critical blow to Hilbert's 'Mathematical Program'; It is a well-known that Hilbert's Mathematical Program sought to base Mathematics (e.g., and by extension also the rest of inductive and deductive Science) upon a logical foundation (e.g., of certain axiomatic definitions); It is also well known that Gödel's Incompleteness Theorem (GIT) has failed Hillbert's Mathematical Program due to its proof that there exists certain 'self-referential' logical-mathematical statements that cannot be determined as "true" or "false" (e.g., or logically 'consistent' or 'inconsistent') from within any hypothetical axiomatic logical-mathematical system... Previously (Bentwich, 2012c,d) it was suggested that perhaps scientific Gödel -like SROCS computational systems may in fact be constrained by the Duality Principle's (generalized) format, thus:

- SROCS: PR{x,y}→['y' or 'not y']/di1...din
- SROCS CR{S,t}→ ['t' or 'not t']/di1...din

wherein it was shown that both inductive ('i') and deductive (ii) SROCS scientific computational systems are necessarily constrained by the Duality Principle (e.g., as part of the broader CUFT). In other words, the Duality Principle's (generalized format) was shown to constrain all (exhaustive hypothetical) Gödel -like (inductive or deductive) scientific SROCS paradigms, thereby pointing at the existence of a singular (higher-ordered) Universal Consciousness Principle ('י') which is solely capable of computing the "simultaneous co-occurrences" of any (exhaustive hypothetical) 'x-y' pairs series. It is important to note, however, that the conceptual computational constraint imposed upon all (Gödel -like) inductive or deductive scientific SROCS paradigms was shown to apply for all of those inductive or deductive (apparent) scientific SROCS paradigms – for which there is an empirically known (or 'knowable') 'x-y' pairs series results!

This latter assertion of the Duality Principle's (generalized proof) may be significant as it both narrows- and emphasized- the scope of the *'scientifically knowable domain'*; In other words, instead of the current 'materialistic-reductionistic' scientific framework which is anchored in a basic (inductive or deductive) SROCS computational format (see above) which inevitably leads to both 'logical inconsistency' and 'computational indeterminacy' that are contradicted by robust empirical findings (e.g., pertaining to the key scientific SROCS paradigms); The Duality Principle (e.g., as one of the postulates within the broader CUFT) proves that the only means for computing the "simultaneous co-occurrences" of any (exhaustive hypothetical) 'x-y' pairs series is carried out by the singular (higher-ordered) Universal Consciousness Principle ('י')... Moreover, the (generalized format of the) Duality Principle goes farther to state that for all other (exhaustive hypothetical) inductive or deductive computational SROCS paradigms – *for which there exists a proven empirical capacity to determine the values of any particular 'x-y' pairs (e.g., empirically "known" or "knowable" 'x-y' pairs results)*- any of these (hypothetical) scientific SROCS computations must be carried out by the CUFT's identified singular Universal Consciousness Principle ('י')!

The (potential) significance of this generalized assertion made by the Computational Unified Field Theory's (CUFT): 'Duality Principle', 'Computational Invariance' principle and Universal Consciousness Principle ('י') is twofold:

a. First, it narrows down the scope of (inductive or deductive) determinable scientific phenomena – to only those (inductive or deductive) 'x-y' relationships for which there is an empirical capacity to determine their "simultaneously co-occurring" values; essentially the *'Universal Consciousness Principle's Computational Program'* anchors itself in the Duality Principle's focus on only those inductive or deductive 'x-y' relationship/s or phenomenon for which there is an empirically 'known' or 'knowable' capacity to determine these 'x-y' pairs values. It is perhaps important to note (in this context) that all of the 'other' inductive or deductive 'x-y' relationship/s which cannot be (empirically) known – "naturally" lie outside the scope of our human (scientific) knowledge (and therefore should not be included, anyway within the scope of Science)... Nevertheless, the strict limitation imposed by the 'Universal Consciousness Principle Computational Program' – may indeed be significant, as it clearly defines the boundaries of "admissible scientific knowledge" to only that scientific knowledge which is based on empirically known or knowable results pertaining to the "simultaneous co-occurrences" of any 'x-y' relationship or phenomenon; (Needless to say that the strict insistence of the Universal Consciousness Computational Program upon dealing only with

b. Second, based on this strict definition of Science as dealing solely with 'empirical knowable' (simultaneously co-occurring) 'x-y' relationship/s or phenomenon – the 'Universal Consciousness Computational Program' may in fact offer a broader alternative to GIT (failing of Hilbert's 'Mathematical Program'); This is because once we accept the Universal Consciousness Principle's Computational Program's (above) strict 'empirical constrains', we are led to the Duality Principle's (generalized) conceptual computational proof that any (exhaustive hypothetical) inductive or deductive scientific SROCS' 'x-y' relationship must be determined by the singular Universal Consciousness Principle ('י') computation of the "simultaneous co-occurrences" of any (exhaustive hypothetical) 'x-y' pairs series. This means that instead of GIT assertion that it is not possible (in principle) to construct a consistent Logical-Mathematical System which will be capable of computing any mathematical (or scientific) claim or theorem, the Universal Consciousness Computational Program asserts that based on a strict definition of Science as dealing solely with empirically knowable 'x-y' relationship/s or phenomenon, we obtain a singular (higher-ordered) Universal Consciousness Principle which is solely responsible for computing the "simultaneous co-occurrences" of any (exhaustive hypothetical) inductive or deductive 'x-y' pairs series (e.g., which were shown by the CUFT to comprise the totality of any single or multiple USCF's frames that are solely produced by this Universal Consciousness Principle). In that sense, it may be said that the Universal Consciousness Principle Computational Program points at the existence of the singular (higher-ordered) Universal Consciousness Principle as constraining- and producing- all inductive or deductive scientific relationship/s or phenomena (e.g., which was also shown earlier and previously to constitute the only "reality" which both produces all USCF's derived secondary computational 'physical properties and also solely exists "in-between" any two such USCF's).

4. THEORETICAL RAMIFICATIONS OF THE UNIVERSAL CONSCIOUSNESS PRINCIPLE

The discovery of the singular Universal Consciousness Principle (alongside its 'Universal Consciousness Computational Program') may bear a few significant theoretical ramifications:

1) **The Sole "Reality" of the Universal Consciousness Principle:** As shown above, all scientific (inductive and deductive) disciplines need to be reformulated based on the recognition that there exists only a singular (higher-ordered) Universal Consciousness Principle ('י') which solely produces- sustains- evolves (and constrains) all (apparent) SROCS (inductive or deductive) 'x-y' relationships; Moreover, this Universal Consciousness Principle is recognized as the sole "reality" that both produces- sustains- and evolves- any of the apparent (four) 'physical' properties of 'space', 'time', 'energy' and 'mass', as well as exists independently of any such 'physical' properties – and is therefore recognized as the only singular "reality", whereas these apparent 'physical' properties are seen as merely 'phenomenal' (secondary computational) manifestations of this singular (higher-ordered) Universal Consciousness Principle "reality".

2) **The Transcendence of 'Material-Causality' by the Universal Consciousness Principle 'A-Causal' Computation:** As shown (above), the acceptance of the Universal Consciousness Principle ('י') as the sole "reality" which both produces- (sustains- and evolves-) all USCF's (secondary computational) 'physical' properties, as well as exists independently "in-between" any (two subsequent) USCF's; (Alongside the Duality Principle's negation of

any apparent SROCS' 'causal' relationships and the 'Computational Invariance' principle indication that only the 'computationally invariant' 'Universal Consciousness Principle' "really" exists whereas the secondary 'computationally variant' physical properties are only 'phenomenal'.) – point at the negation of any "real" material-causal ('x-y') relationships, but instead indicate that there can only exist a singular (higher-ordered) Universal Consciousness Principle 'a-causal' computation of the "simultaneous co-occurrences" of any exhaustive hypothetical inductive or deductive 'x-y' pairs' series... (As shown earlier, the strict negation of the existence of any "real" 'material-causal' 'x→y' relationships was evinced by the simple fact that according to the CUFT's model there cannot exist any "real" computationally variant 'physical' or 'material' property that can "pass" across any two subsequent USCF's, but only the computationally invariant "real" Universal Consciousness Principle which exists singularly – as solely producing all apparent secondary computational 'physical' properties as well as existing independently "in-between" any two such subsequent USCF's frames.) Indeed, the need to replace all apparent 'material-causal' 'x-y' SROCS relationships by a singular (higher-ordered) Universal Consciousness Principle computation of the 'simultaneous co-occurrences' of all possible inductive or deductive 'x-y' pairs series was shown to apply to all of the key (apparent) scientific SROCS paradigms (including: Darwin's Natural Selection Principle and associated Genetic Encoding hypothesis, Neuroscience's Psychophysical Problem of human Consciousness as well as to all Gödel-like hypothetical inductive or deductive SROCS paradigms; what this implies is that for all of these apparent SROCS scientific paradigms the sole "reality" of the Universal Consciousness Principle forces us to transcend each of the (particular) 'material-causal' x-y relationships in favor of the Universal Consciousness Principle's singular computation of all (exhaustive hypothetical) 'x-y' pairs series; Thus, for example, instead of Darwin's current 'Natural Selection Principle' SROCS material-causality thesis, which assumes that it is the direct (or indirect) physical interaction between the organism and its Environmental Factors that causes that organism to 'survive' or be 'extinct', the adoption of the Universal Consciousness Principle (and Duality Principle) postulates brigs about a recognition that there is only a singular (Universal Consciousness based) conceptually higher-ordered 'a-causal' computation of the "simultaneous co-occurrences" of an exhaustive hypothetical pairs series of 'organism' and 'Environmental Factors' (e.g., which are computed as part of the Universal Consciousness Principle's production of the series of USCF's frames).

3) **Possible Resolution of Physical (and Mathematical) Conundrums:** It is suggested that certain key Physical (and Mathematical) Conundrums including: Physics "dark energy", "dark matter" and "arrow of time" enigmas may be potentially resolved through the application of this singular 'Universal Consciousness Principle'; this is because according to the CUFT, all (four) 'physical' properties of 'space', 'time', 'energy' and 'mass' are (in reality) solely produced by the Universal Consciousness Principle (e.g., as secondary computational 'phenomenal' properties); Hence, the key enigma of "dark energy" and "dark matter" (e.g., the fact that based on the calculation of the totality of 'mass' and 'energy' in the observable cosmos the expansion of the universe should not be as rapid as is observed – which is currently interpreted as indicating that approximately 70-90% of the "energy" and "mass" in the universe in "dark", that is not yet observable) – may be explainable based on the CUFT's delineation of the Universal Consciousness Principle's (extremely rapid) computation of the series of USCF's. This is due to the fact that according to the Universal Consciousness Principle's (previously discovered: Bentwich, 2012a) 'Universal Computational Formula' the production of any "mass" or "energy" ("space" or "time") 'physical' properties – are entirely (and solely) produced through the Universal Consciousness Principle's computation of the degree of 'Consistency' (e.g., 'consistent' or 'inconsistent') across two other Computational Dimensions, i.e.,

'Framework' ('frame' vs. 'object') and 'Locus' ('global' vs. 'local'): Thus, for instance it was shown that any "mass" measurement of any object in the universe is computed by the Universal Consciousness Principle ('י') as the degree of 'consistent-object' measurement (of that particular) object across a series of USCF frames.

Hence, by extension, the totality of the "mass" measured across the entire physical universe should be a measure of the degree of consistent-object/s values across a series of USCF's! Note, however, that based on the abovementioned recognition that in "reality" – only the Universal Consciousness Principle ('י') "exists" (e.g., both as producing any of the USCF's derived four secondary computational 'physical' properties as well as existing independently "in-between" any two such USCF's frames), and therefore that only this Universal Consciousness Principle "really" produces all of the (apparent) "mass" and "energy" in the 'physical' universe (e.g., rather than the "energy" and "mass" in the 'physical' universe being "caused" by the "material" objects in the cosmos)... Hence, also all of the "energy" in the physical universe is solely produced by this (singular) Universal Consciousness Principle, e.g., as a measure of the degree of 'inconsistent-frame' (changes) of all of the objects (I the universe) across a series of USCF's frames. Therefore, according to the CUFT, the explanation of all of the "mass" and "energy" values observed in the 'physical' universe – should be solely attributed to the operation of the Universal Computational Principle, i.e., through its (extremely rapid) computation of the rapid series of USCF's (respective secondary computational measures of the abovementioned degree of 'consistent-object': "mass", or 'inconsistent-frame': "energy"). We therefore obtain that the (accelerated) rate of expansion of the physical universe – should be explained (according to the CUFT) based on the Universal Consciousness (extremely rapid) computation of the USCF's (e.g., which gives rise to the apparent secondary computational 'physical' measures of 'consistent-object': "mass" or 'inconsistent-frame': "energy"), rather than arise from any 'material-causal' effects of any (strictly hypothetical) "dark mass" or "dark energy"... (Once again, it may be worth pointing at the abovementioned conceptual computational proof that there cannot be any transference of any "physical" property (entity or effect etc.) across any (two subsequent)

USCF's frames, but only the retention- or evolution- of all of the spatial pixels' "physical" properties by the singular Universal Consciousness Principle across the series of USCF's – which therefore also precludes the possibility of any "real" "material" effects exerted by any "dark" mass or energy on the expansion of the 'physical' universe across a series of USCF frames.)

Similarly, the "arrow of time" conundrum in modern Physics, e.g., which essentially points at the fact that according to the laws of Physics, there should not be any difference between the physical pathways of say the "breaking of a glass cup into a (thousand) small glass' pieces" and the "re-integration of these thousand glass' pieces into a unitary glass cup"! In other words, according to the strict laws of Physics, there should not be any preference for us seeing "glasses" break into a thousand pieces – over our seeing of the thousand pieces become "reintegrated" into whole glass cups (again), which is obviously contradicted by our (everyday) phenomenal experiences (as well as by our empirical scientific observations)... Hence, according to the current state of (quantum and relativistic) models of Physical reality – there is no reasonable explanation for this "arrow of time" apparent empirical "preference" for the "glass breaking into pieces" scenario over the "reintegration of the glass pieces" scenario...

However, it is suggested that according to one of the CUFT critical empirical predictions (previously outlined: Bentwich, 2012b) this "arrow of time" Physical conundrum may be resolved: This is because one (of three) critical empirical predictions of the CUFT assert the possibility of reversing any spatial-temporal sequence associated with any given 'electromagnetic spatial pixel' through the appropriate manipulation of that object's (or event's) electromagnetic spatial pixel values (across a series of USCF's): It was thus indicated that if we were to accurately record the spatial electromagnetic pixels' values of any particular object (e.g., such as an amoeba or any other living organism for instance) across a series of USCF's frames (e.g., or even through a certain sampling from a series of USCF's), and to the extent that we could appropriately manipulate these various electromagnetic spatial pixels' values in such a manner which allows us to reproduce that objects' electromagnetic spatial pixels' values (across the measured series of USCF's) – in the reversed spatial-temporal sequence, then it may be possible to reverse the "flow of time" (e.g., spatial-temporal electromagnetic pixels' sequence). In this way it should be possible (according to one of the critical predictions of the CUFT) to actually "reverse" the "arrow of time" (e.g., at least for particular object/s or event/s: such as for instance, bring about a situation in which a "broken glass cup may in fact be reintegrated"...)

REFERENCES

1. J Bentwich, 2003aFrom Cartesian Logical-Empiricism to the'Cognitive Reality': A Paradigmatic Shift, *Proceedings of Inscriptions in the Sand, Sixth International Literature and Humanities Conference*, Cyprus

2. J Bentwich, 2003bThe Duality Principle's resolution of the Evolutionary Natural Selection Principle; The Cognitive 'Duality Principle': A resolution of the 'Liar Paradox' and 'Gödel's Incompleteness Theorem' conundrums; From Cartesian Logical-

3. Empiricism to the 'Cognitive Reality: A paradigmatic shift *Proceedings of 12th International Congress of Logic, Methodology and Philosophy of Science,* August Oviedo, Spain

4. J Bentwich, 2003cThe cognitive'Duality Principle': a resolution of the'Liar Paradox' and'Gödel's Incompleteness Theorem' conundrums,*Proceedings of Logic Colloquium,* Helsinki, Finland, August 2003

5. J Bentwich, 2004The Cognitive Duality Principle: A resolution of major scientific conundrums, *Proceedings of The international Interdisciplinary Conference,* Calcutta, January

6. J Bentwich, 2006The 'Duality Principle': Irreducibility of sub-threshold psychophysical computation to neuronal brain activation. *Synthese*, 153 3 451 455

7. J Bentwich, 2012aQuantum Mechanics / Book 1 (979-9-53307-377-3Chapter title: The'Computational Unified Field Theory' (CUFT): Harmonizing Quantum Mechanics and Relativity Theory.

8. J Bentwich, 2012bQuantum Mechanics / Book 1 (979-9-53307-377-3Chapter 23, 551 598Theoretical Validation of the Computational Unified Field Theory.

9. J Bentwich, 2012cQuantum Mechanics (In Press) Chapter title: The'Computational Unified Field Theory' (CUFT): A Candidate Theory of Everything.

10. M Born, 1954The statistical interpretation of quantum mechanics, *Nobel Lecture, December 11, 1954*

11. G Brumfiel, 2006Our Universe: Outrageous fortune. *Nature*, 439 10 12

12. C Darwin, 1859 *On the Origin of Species by Means of Natural Selection, or the Preservation of Favoured Races in the Struggle for Life* John Murray, London; modern reprint Charles Darwin, Julian Huxley (2003). *On The Origin of Species*. Signet Classics. 0-45152-906-5

13. J Ellis, 1986The Superstring: Theory of Everything, or of Nothing? *Nature, 23* 6089 595 598

14. B Greene, 2003 *The Elegant Universe,* Vintage Books, New York

9

ON THE DUAL CONCEPTS OF 'QUANTUM STATE' AND 'QUANTUM PROCESS'

Cynthia Kolb Whitney

Galilean Electrodynamics, USA

1. INTRODUCTION

Many of us who made a living in the 20th century did so by functioning as some kind of engineer. Though schooled mostly in Physics, this author often functioned in those days as an engineer. It was a good continuing education. One aspect of it was the big tool kit in use. For example, some subject systems were best viewed in the frequency domain: a system functioning as a filter would suppress some frequencies and enhance others. But other systems were better viewed in the time domain: a system functioning as a controller would take a time series of input signals and produce a time series of output commands. Neither approach was considered more right, or more fundamental, than the other. They were complementary.

But in the 20th century, things felt less eclectic in Physics. Especially in the literature of Quantum Mechanics (QM), there often seemed to be a lot of passion about what viewpoint was allowed, and what viewpoint was not allowed. We were taught that it just was not correct to think of an atom as a nucleus with electrons in orbits around it. There could *not* be orbits; there *had to* be only 'orbitals', a new word coined to refer to complex wave functions that extended over all space, and provided only spatial densities of probability, in the form of squared amplitude. Except for its phase factor, there was no sense of time-line to an orbital. It was a stable state.

So in QM, the emphasis was all on the stable states. Between the stable states, there could occur transitions, resulting in emission or absorption of a photon, but the state transitions themselves were essentially instantaneous, and not open to study. This emphasis on the stable states, and the avoidance of the transitions between them, implied that questions about the details of state transitions should be regarded as illegitimate.

Back at the turn of the 20th century, there was a good reason for the avoidance of details about process in QM: we did not understand how any atom could resist one totally destructive process that was expected within the context of classical electrodynamics. Even the simplest atom, the Hydrogen atom, was expected to continuously radiate away its orbit energy, and so quickly collapse. So ever paying attention to any details of process looked fraught with peril.

A way of escaping the issue of process in QM came along with the discovery of the photon. Quantization of light was implied in the spectrum of blackbody radiation, and demonstrated in the photoelectric effect. Those developments gave us Planck's constant. Planck's constant provided a constraint for defining the ground state of the Hydrogen atom. So Schrödinger wrote it into his equation for the wave function of the electron in the Hydrogen atom. The Schrödinger equation produced a set of solutions representing a whole family of stationary states for the Hydrogen atom. We could forget process, and focus on those stationary states.

But today, one of the big application areas for QM is in development of computational approaches fitting the name Quantum Chemistry (QC). Chemistry is largely about reactions, and certainly every reaction is a complex process. So the chemical reactions are like the quantum state transitions: they have, not only the stationary state before, and the stationary state after, but also something of interest in between. So the historical injunction against inquiry into the specifics of quantum state transitions tends to inhibit the full application of QM to the process-related problems that QC presents. So consideration of process is no longer avoidable for QM.

Fortunately, history is never the final story; it exists mainly to be updated from time to time. A Chapter in an earlier Book in this series (Whitney, 2012) argued that our understanding of Maxwell's Electromagnetic Theory (EMT) at the turn of the 20th century was incomplete. When we develop a description for the photon based on EMT, we learn some facts that have bearing on the communication between the electron and the proton in the Hydrogen atom, and how that communication in turn supports the continuing existence of the atom as a system. So consideration of a quantum process becomes less perilous for QM.

But the practical difficulties are numerous. In many cases, molecules involve numbers of atoms that are too small for any kind of statistical ideas to be applicable, but too large for traditional QM calculations to be practical; hence, they are altogether awkward to address. Furthermore, Chemistry is all about reactions, which can involve many molecules. And sometimes there are multiple reaction steps, or even multiple paths, each one with multiple steps, each one with a time line worthy of detailed numerical study.

The practical difficulties of QC arise largely because the most common way of thinking about QM is still in terms of wave functions. Their amplitudes are squared to make probability density functions, multiplied by functions or differential operators representing variables of interest, and integrated over argument variables. It can add up to way too much computation.

So how can QM better meet the needs of QC? A potentially helpful concept comes out of QM: the concept of duality. Abad and Huichalaf (2012) describe it in terms of *seeming* contradiction, and*seeming* is certainly the right characterization, for duality is *not* really a contradiction at all. Consider, for example, the traditional wave *vs.* particle duality of light. The earlier Chapter (Whitney, 2012) presented a model for the photon based on Maxwell's four coupled field equations, together with boundary conditions representing the source and the receiver of the photon. The Maxwell photon model is pulse-like at emission, and evolves into an extended wave-like condition, and then collects back into a more confined pulse-like condition for absorption. In QM, *all* observable objects are like the Maxwell photon model, in that they present both particle-like and wave-like aspects. Which aspect is seen just depends on when and how one observes the object.

To meet the needs of QC, we clearly need to develop and exploit another duality analogous to the traditional wave *vs*. particle duality of observed objects. We need a duality of observer descriptions: state descriptions *vs*. process descriptions. Where the traditional QM approach starts with the idea of quantum state, which is something stationary, the dual approach must start with the idea of quantum process, which naturally has a time line to it. The temporal evolution of the Maxwell photon model is fundamentally a process, and it can be followed in detail, with no untoward disasters, so it is a promising point of departure for this work.

One key thing about photons is that they have finite energy. In this respect, photons are completely different from infinite plane waves, which have infinite energy. The Maxwell photon model has the finite total energy as needed. That energy is always trapped in the space between the source and the receiver. So another key role in the photon model is played by mathematical boundary conditions. Section 2 picks up where the earlier work left off, discussing in more detail what the boundary conditions are, what they do, and how they do it.

Photons were not known in Maxwell's day, so the implication of their finite energy was not then appreciated. Coupled with their finite propagation speed, their finite energy causes the definite Arrow of Time that has long been considered such a mystery in Physics. Many people suppose that the Arrow of Time has to do with Thermodynamics, because that subject deals with entropy and irreversibility. Searching for a mechanism, many people would think of friction in Newtonian mechanics. When told that Electrodynamics displays the Arrow of Time, many people think first of the friction-like effect of radiation reaction acting on accelerating charges. But actually, the Arrow of Time is present quite apart from anything that happens to material particles. It appears in the photon itself.

There is a reason why this fact was not emphasized a long time ago. Section 3 recalls how Maxwell's coupled field equations were immediately inserted, one into another, in order to reduce the set of four coupled field equations into a set of two un-coupled wave equations. The two un-coupled wave equations clearly display the finite propagation speed, but they totally hide the effect of finite energy.

The problem is this: the two un-coupled wave equations are less restrictive than the four coupled field equations are, so they have a larger set of solution functions, some of which do *not* also solve the four coupled field equations. One example of such a solution is a finite-energy pair of orthogonal E and B pulses that travels without distortion and faithfully delivers information.

In the early 20th century, Einstein used this kind of solution for the role of 'signal' in his Special Relativity Theory (SRT). His goal was to capture the spirit of Maxwell's EMT into SRT. But his signal model is inadequate for that job. The Maxwell photon model better captures the spirit of Maxwell, and only slightly modifies the SRT results, and better ties those results into QM.

What about the QM of material particles? Section 4 revisits the Schrödinger equation. Viewed as the analog to a statement from Newtonian mechanics, it has no essential irreversibility to it. That is one reason why its solutions are *stable* states. That is why something else is needed for the study of transitions between the quantum states of atoms, and for the numerous chemical processes that QC should address.

Section 5 points out that the main thing for Chemistry is not revealed in the Hydrogen atom, due to the fact that the Hydrogen atom has only one electron. The main thing for Chemistry is the variety of relationships among *multiple* electrons. Chemistry data suggest that electrons can sometimes actually attract, rather than repel, each other. That propensity can be important in driving chemical reactions. The calculation approach suitable for QC was introduced in the earlier Chapter (Whitney 2012). It is called Algebraic Chemistry (AC). It is a big subject, detailed further in a full Book (Whitney 2013).

Section 6 concludes this Chapter. It draws a lesson from all the problems treated here. The lesson is that we sometimes actually make problems very much worse than they need to be by oversimplifying them. Many seeming mysteries in physical science are nothing but our own creations.

2. MORE ABOUT PHOTONS

The earlier Chapter (Whitney, 2012) revisited the quantum of light, the photon, and its relationship to Classical Electrodynamics. That Chapter argued that a simple mental picture of a photon as a pair of electric and magnetic field pulses that travel together, but do not change their pulse shapes, does not comport with Maxwell's four coupled field equations. Instead, there has to be a temporal evolution, first from pulses emitted by a source, into a waveform shape extended in the propagation direction, then back to a more compact shape, concluding with absorption by a receiver; in short, a whole time-line *process*.

This Section gives further mathematical detail about the temporal evolution of the photon waveform. In summary, the evolution begins with emission of Gaussian field pulses at the photon source. After emission, the fields develop according to Maxwell's coupled field equations. The development is constrained and guided by boundary conditions that represent the initial source of the photon and the ultimate receiver of the photon. In the end, all the energy accumulates near the receiver, and can finally be swallowed by it.

In more detail, Maxwell's coupled field equations cause the Gaussians in E and B to beget first-order Hermite polynomials in B and E, and then those beget second-order Hermite polynomials in E and B, and so on, indefinitely as time goes on. The roots of each newly generated Hermite polynomial interleave with the roots of the previously generated Hermite polynomial, with one more root being added at each step of the process. This process is illustrated with Figure 1 in (Whitney, 2012). It shows spreading of the waveform in its propagation direction.

Next, there have to be boundary conditions to represent the source and the receiver. The boundary conditions can be like those representing the mirrors in a laser cavity: they can enforce a zero in the E field at the boundary locations. As a result of the zero-E condition at the source, the spreading never causes any backflow of field energy into the space behind the source. And as a result of the zero-E condition at the receiver, the waveform spreading never causes any field energy to propagate into the space beyond the receiver. So eventually all the energy just 'piles up' before its final absorption into the receiver.

A mental picture of the photon in terms of electric and magnetic fields is quite complicated, even without the boundary conditions. There have to be two field vectors EE and BB in orthogonal directions to create a cross product, the Poynting vector of energy propagation. And there have to be two such pairs of fields, one a quarter cycle out of phase with the other in both space and time, to create circular polarization. That makes a quartet of field vectors to think about. Then to create the boundary conditions, there have to be two more such quartets of field vectors, arranged to propagate in the opposite direction, and placed to provide the E-field cancellations at the boundaries. But each of these field vector quartets slightly spoils the boundary condition fixed by the other one. So then an infinite regression of more and more field quartets is demanded.

A mental picture of the photon can be formulated much more simply in terms of its overall profile of energy density, (E2+B2)/2(E2+B2)/2. Before the boundary conditions are imposed, the energy density profile is always a simple Gaussian, the height of which decreases over time, and the width of which increases over time. Then to impose the boundary conditions, the infinite Gaussian tails get folded, and refolded infinitely many times over, back into the space between the source and the receiver. That means the energy density profile is a slightly deformed from Gaussian in the vicinity of the source, and the vicinity of the receiver: cut off sharply at those points, and slightly more than doubled in height near them, because zero EE means double BB and double energy density. Of course, the total energy, the integral of the energy density profile between the source and the receiver, never changes.

Figure 1 illustrates this mental picture of the photon as a changing energy density profile with constant total energy. The three data series plotted correspond to the energy density profiles near the beginning, in the middle, and near the end, of the photon propagation scenario. Take note of the phenomenon of waveform spreading. It is not reversible. It shows the Arrow of Time.

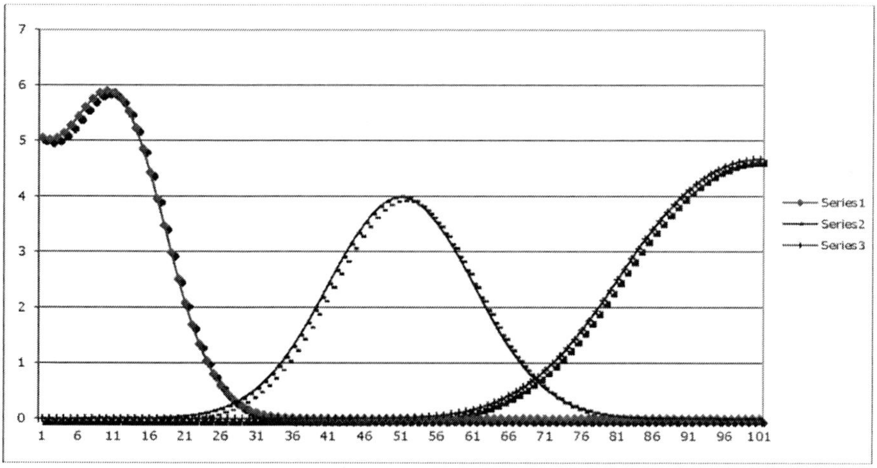

Figure 1. Photon energy density profile at three stages.

3. ON THE ARROW OF TIME

There is an interesting history to the Arrow of Time, and it begins with the beginning of Mathematical Physics itself. Newton developed his laws of mechanics at a time of great interest in celestial events. In that sphere, the absence of contact invites the assumption of no friction, and in the absence of friction, Newtonian mechanics is invariant under time reversal. In this case, there is in principle no Arrow of Time. However, in practice there is a related problem about chaos, which develops in the mathematics when three or more bodies interact. It was certainly impossible to solve for three trajectories in closed form, *i.e.* as simple functions of time. Even today, it is not entirely clear exactly what *is* possible to do.

Maxwell developed his electromagnetic theory at a time of developing industrial technology. In that domain, the first thing encountered is limitation. With real-world machines, there is always an Arrow of Time, and it is best described with the science of Thermodynamics. In Maxwell's electrodynamics, there are two important features. First, there is *not* instantaneous-action-at-a-distance; there is a finite propagation speed. It shows up clearly with propagating waves. Second, there can be infinite plane waves in the mathematics, but not in physical reality. Reality is *always* about finite energy.

The finite propagation speed was discovered first. People quickly transformed Maxwell's four coupled field equations into something more familiar: two un-coupled wave equations. They then often fixated on the infinite plane wave solutions of those equations. The implications of finite energy were then left for investigation later. The job was partially addressed in the study of Optics: finite apertures create diffraction patterns transverse to the propagation direction. But the implications of pulsing in the propagation direction itself would become interesting only much later; for example, with the invention of pulsed lasers.

But much work was left undone at that time. So let us pursue the investigation a little further now. In modern notation and Gaussian units, Maxwell's four coupled field equations go

$$\nabla \cdot B=0,\ \nabla \cdot D=4\pi\rho,\ \nabla \times E+\frac{1}{c}\partial B\Big/\partial t=0,\ \nabla \times H-\frac{1}{c}\partial D\Big/\partial t=\frac{4\pi}{c}J.$$

Here BB is the magnetic field vector and EE is the electric field vector. In free space $H=B/\mu_0,\ 1/c=\sqrt{\varepsilon_0\mu_0}$, and charge density ρρ and current density vector JJ are zero.

The two uncoupled wave equations go

$$\nabla^2 E-\frac{1}{c^2}\partial^2 E\Big/\partial t^2=0,\ \nabla^2 B-\frac{1}{c^2}\partial^2 B\Big/\partial t^2=0.$$

where 1/c2=ε0μ01/c2=ε0μ0.

Because the time derivatives in the two un-coupled wave equations are second order, the sign attributed to time cancels out. So the two un-coupled wave equations are invariant under time reversal. This fact means the two un-coupled wave equations are *not equivalent* to the four coupled field equations. The two un-coupled wave equations have a *larger set of solutions* than do the four coupled field equations. So while *all* of the solutions to the four coupled field equations will satisfy the two un-coupled wave equations, only *some* of the solutions to the two un-coupled wave equations will satisfy the four coupled field equations.

So some solutions to the two un-coupled wave equations *do not satisfy* the four coupled field equations. The infinite plane wave is one of those. It has the parameter cc, which was interpreted as the speed of propagation for light, not just in the context of the infinite plane wave, but also in every other context; in particular, in the context of the concept of *signal* that Einstein used in the development of Special Relativity Theory (SRT).

But the signal for SRT cannot be just an infinite plane wave. A signal has to have some discernable feature in order to carry any information. Bracken (2012) discusses the mathematical concept of information, and its relation to entropy. The infinite plane wave has a delta-function spectrum, meaning just one wavelength, zero entropy, and hence zero information content.

Instead of an infinite plane wave, a signal needs finite E and B pulses. These can be constructed from a spectrum of infinite plane waves with infinitesimal amplitudes. But as we have seen, such pulses have to evolve over time into spread-out waveforms. Pulses without such evolution can satisfy the two uncoupled wave equations, but they *cannot* satisfy Maxwell's four coupled field equations.

The purpose of the SRT signal concept was to capture the essence of Maxwell's EMT into SRT. So in retrospect, it does not seem very appropriate to have used a model for the signal that does not satisfy Maxwell's four coupled field equations. It seems more appropriate to use the Maxwell photon model instead. Einstein did develop SRT at nearly the same time as he did his Nobel-prize winning work about the photoelectric effect – and the photon. But mysteriously, the idea that the SRT signal is similar to the photon did not come up at that time.

Better late than never, we can tackle that problem now. Clearly, at the beginning of the propagation scenario, the boundary condition at the source dominates the energy profile, and at the end of the propagation scenario, the boundary condition at the receiver dominates the energy profile. From this assessment, it is easy to imagine a generalization of the propagation scenario. Like the mirrors in a tunable laser, the source and the receiver in the propagation scenario can be imagined to move relative to each other. This possibility then suggests the deep question:

What is the frame of reference for light speed *c* ?

This question lies at the heart of SRT. There, the reference for light speed cc is any and all observers. That is Einstein's Second Postulate. To be fair, it was also the hidden Assumption of all prior works in Electrodynamics. So Einstein is to be credited for making it so explicit.

But a Postulate is a rather formalized way of stating an Assumption, and, as such, does not invite as much scrutiny, or convey as much reward when scrutiny is given, as an Assumption would do. Linguistically, the Second Postulate may seem quite incomprehensible, but its status of Postulate, rather than Assumption, invites a level of deference. This appears to be a partial answer for the 'Why?' articulated above.

However, use of the Maxwell photon model instead of the Einstein signal does allow a more linguistically normal kind of statement. Observe that during times soon after the emission, the boundary condition that is numerically more significant is the one demanding zero E at the source. But during times just before the reception, the boundary condition that is numerically more significant is the one demanding zero E at the receiver. The consequence is that the photon energy starts out traveling at speed cc relative to the source, but then finishes up traveling at speed cc relative to the receiver. In between, the reference for cc has to transition gradually from one frame to the other, along with the numerical dominance of the corresponding boundary conditions.

This analysis leaves us with a conundrum: If the Maxwell photon model can be adopted as the signal for developing SRT, then the Einstein's Second Postulate is true at the moment of reception, but not generally true at any moment before that event. So then SRT *cannot* be universally valid. But if the Maxwell photon model *cannot* be adopted as the signal for developing SRT, then SRT should not be considered as founded in Maxwell's EMT. So then SRT is not founded in *any* prior science. This conundrum does not disqualify SRT as a useful theory, but it does mean that prudence demands investigation of other theories too.

Use of the Maxwell photon model instead of Einstein's Second Postulate does produce results somewhat different from those of Einstein's SRT. Fortunately, it is not very difficult to work out these slightly different results. The different results that are important for connecting better with QM are given in Whitney (2012). In summary, they are the following: 1) Within the Hydrogen atom, there exists not just the one electromagnetic process, the radiation from the accelerating electron; there exist*three* electromagnetic processes. The other two are: 2) torquing internal to the atomic system, and 3) circular motion of its center of mass. The torquing produces an energy gain mechanism that more than compensates the energy loss due to radiation. But the center-of-mass motion amplifies the radiation loss, bringing the atomic system into its final balance. The combination of these *three* electromagnetic processes, instead of just one, makes it possible to model a stable Hydrogen atom electromagnetically.

Here is how finite signal propagation speed causes the second and third effects. Each particle is attracted to a former position of the other. The forces are not central, and are not even balanced. The non-centrality centrality causes the torquing. The force imbalance causes the center of mass circulation. That in turn causes Thomas rotation, which amplifies the radiation. It is interesting that Thomas rotation was first discovered in the context of sequential Lorentz transformations, and is generally believed to be a consequence of SRT. See, for example, De Zela (2012). But Thomas rotation actually does *not* depend on Lorentz transformations; it emerges just as well from sequential Galilean transformations. It is not a relativistic effect.

4. MORE ABOUT THE HYDROGEN ATOM

The treatment of the Hydrogen atom given in Whitney 2012 was the first-order approximation, in which the cosine and the sine of an angle traversed around the circular orbit were represented to first order, by unity and by the angle itself. Let us now go further. The total radiation power changes from

$$P_{\text{total radiated}} \approx 2^4(2e^2/3c^3)a_e^2 = (2^5e^6/m_e^2)/3c^3(r_e+r_p)^4$$

to

$$P_{\text{total radiated}} \approx \left[(2^5e^6/m_e^2)/3c^3(r_e+r_p)^4\right] \times \cos^2(v_e/2c).$$

The total torquing power changes from

$$P_{\text{torquing}} = \left[(e^4/m_p)/c(r_e+r_p)^3\right] \times \frac{1}{2}\left[\frac{\sin(v_e/2c)}{(v_e/2c)} + \frac{\sin(v_p/2c)}{(v_p/2c)}\right].$$

to

$$P_{\text{torquing}} = \left[(e^4/m_p)/c(r_e+r_p)^3\right] \times \frac{1}{2}\left[\frac{\sin(v_e/2c)}{(v_e/2c)} + \frac{\sin(v_p/2c)}{(v_p/2c)}\right].$$

Observe that both of these expressions are oscillatory. But where $P_{\text{total radiated}}$ is confined to non-negative values, P_{torquing} is not. So while $P_{\text{total radiated}}$ always represents a mechanism for energy loss,

P_{torquing} does *not* always represent a mechanism for energy gain. For electron orbit radius rere below the Hydrogen ground state orbit,

P_{torquing} oscillates between giving energy gain and giving energy *loss*. Therefore, there exist many values of rere where $P_{\text{total radiated}} = P_{\text{torquing}}$, and the Hydrogen atom can exist, and even persist. These represent 'sub-states' of the Hydrogen atom. Such states are sometimes discussed in the literature of experimental physics, but never in the literature of theoretical physics. For one thing, they seem to involve orbit speeds beyond the speed of light, and so violate SRT. For another thing, they are not within the universe of discourse of the Schrödinger approach to the Hydrogen atom.

The Schrödinger approach to the hydrogen atom

Schrödinger's famous equation representing a particle, such as an electron, reads:

$$i\hbar\frac{\partial}{\partial t}\Psi(r,\,t)=\left[-\frac{\hbar^2}{2m}\nabla^2+V(r,\,t)\right]\Psi(r,\,t),$$

where Ψ(r,t)Ψ(r,t) is the wave function, rr is position in three-dimensional space, and tt is time. The *ħħ* is the reduced Planck's constant h/2πh/2π, mm is the particle mass, ∇2∇2 is the usual three-space second-derivative operator, and V(r,t)V(r,t) is the potential energy, created for example by a nucleus. For an atom that is not moving, and is not perturbed by some measuring device, V(r,t)V(r,t) is time-invariant, reducing to V(r)V(r).

Schrödinger's equation is about a wave function, and not about a particle. So in the beginning, it seemed to have no clear foundation in the science prior to its own time. It was taken as a gift from heaven. But actually, Schrödinger's equation *does* have a foundation – just not entirely within Physics, but also partly within Engineering Science. The following analysis shows that Schrödinger's equation reduces to a classical equation based on Newton's laws. The reduction uses Fourier transforms, a tool very commonly used in Engineering Science.

Consider first the wave function Ψ(r,t)Ψ(r,t). By definition, it satisfies the normalization condition

$$\int_{-\infty}^{\infty}\Psi(r,\,t)\,\Psi^*(r,\,t)d^3r=\int_{-\infty}^{\infty}|\,\Psi(r,\,t)|^2\,d^3r=1.$$

The function Ψ(r,t) has four-dimensional Fourier transform Ψ(p,E), where pp is momentum and EEis total energy. This Fourier transform is defined by:

$$\Psi(p,\ E)=\frac{1}{(2\pi)^2}\int_{-\infty}^{\infty}\Psi(r,\ t)\exp[i(p\cdot r-E\cdot t)/\hbar]d^{3}r\,dt.$$

This function too satisfies a normalization condition

$$\int_{-\infty}^{\infty}\Psi(p,\ E)\,\Psi^{*}(p,\ E)d^{3}p=\int_{-\infty}^{\infty}|\,\Psi(p,\ E)\,|^{2}\,d^{3}p=1.$$

The corresponding *inverse* Fourier transform is defined by

$$\Psi(r,\ t)=\frac{1}{(2\pi)^2}\int_{-\infty}^{\infty}\Psi(p,\ E)\exp[-i(p\cdot r-E\cdot t)/\hbar]d^{3}p\,dE.$$

(An aside: definitions of Fourier transforms for other applications sometimes deploy the factors of 2π2πdifferently, although always such that the round trip from one space to the other and back again has 1/2π1/2π for each dimension.)

Observe first of all that if Ψ(r,t)Ψ(r,t) is very sharply peaked over a small range ΔrΔr, say centered at r=0r=0, then Ψ(p,E)Ψ(p,E) will be very spread out over a large range ΔpΔp, centered at p=0p=0. And *vice versa*: large ΔrΔr makes for small ΔpΔp. That means the Fourier pair of functions Ψ(r,t)Ψ(r,t) and Ψ(p,E)Ψ(p,E)automatically generates a relationship that looks like the Heisenberg uncertainty relationship.

The product Δr·ΔpΔr·Δp has its minimum possible value when Ψ(r,t)Ψ is a Gaussian function, in which case Ψ(p,E) is also a Gaussian function. For the Physics application, the product (Δr·Δp)Gaussian Planck's constant hh.

There is no way around the $\Delta r\cdot\Delta p\geq(\Delta r\cdot\Delta p)_{\text{Gaussian}}$. It is a property of Mathematics in general, and not of QM in particular. But physicists do worry about its meaning for QM. For example,Cini (2012) seems ready to remove the deBroglie and Schrödinger classical probability wave approaches from the main narrative of QM, and begin it instead with quantum field theory. One problem with this strategy is the risk of putting too much trust in SRT, which appears possibly flawed in its founding Postulate.

In contrast to physicists, engineers just accept the law $\Delta r \cdot \Delta p \geq (\Delta r \cdot \Delta p)_{\text{Gaussian}}$ because in their world there *never* exists a measurement without a spread, and they regard *any* proposed perfectly precise physical quantity as just a metaphysical idea, and not a real physical thing.

The present analysis proceeds in that spirit. The next step is to rewrite Schrödinger's equation in the form:

$$\left[-\frac{\hbar^2}{2m}\nabla^2\Psi(r,t)\right]\Psi^*(r,t)+[V(r,t)\Psi(r,t)]\Psi^*(r,t)=\left[i\hbar\frac{\partial}{\partial t}\Psi(r,t)\right]\Psi^*(r,t).$$

With its seemingly superfluous Ψ∗(r,t)Ψ*(r,t) factors, this form of Schrödinger's equation looks more complicated than the original form. But this form ultimately leads to tremendous simplification and explanatory power. And, as in Berkdemir (2012), the objective here is mainly pedagogical.

The first term on the left side of the rewritten Schrödinger's equation is:

$$\left[-\frac{\hbar^2}{2m}\nabla^2\Psi(r,t)\right]\Psi^*(r,t)=$$

$$\frac{1}{(2\pi)^2}\int_{-\infty}^{\infty}\frac{p^2}{2m}\Psi(p,E)\exp[-i(p\cdot r-E\cdot t)/\hbar]d^3p\,dE\frac{1}{(2\pi)^2}\int_{-\infty}^{\infty}\Psi^*(p',E')\exp[+i(p'\cdot r-E'\cdot t)/\hbar]d^3p'\,dE'$$

$$=\int_{-\infty}^{\infty}\frac{p^2}{2m}\Psi(p,E)\Psi^*(p,E)d^3p\,dE=\left\langle\frac{p^2}{2m}\right\rangle$$

where ⟨⟩⟨⟩ indicates statistically average value.

The second term on the left side is just:

$$\int_{-\infty}^{\infty}V(r,t)\Psi(r,t)\Psi^*(r,t)d^3r\,dt=\langle V(r,t)\rangle$$

The one term on the right side is:

$$\left[i\hbar\frac{\partial}{\partial t}\Psi(r,t)\right]\Psi^*(r,t)=$$

$$\frac{1}{(2\pi)^2}\int_{-\infty}^{\infty}E\,\Psi(p,E)\exp[-i(p\cdot r-E\cdot t)/\hbar]d^3p\,dE\frac{1}{(2\pi)^2}\int_{-\infty}^{\infty}\Psi^*(p',E')\exp[+i(p'\cdot r-E\cdot t)/\hbar]d^3p'\,dE'$$

$$=\int_{-\infty}^{\infty}E\,\Psi(p,E)\Psi^*(p,E)d^3p\,dE=\langle E\rangle$$

So viewed in this way, Schrödinger's equation reads:

$$\left\langle \frac{p^2}{2m} \right\rangle + \langle V(r, t) \rangle = \langle E \rangle.$$

This presentation of Schrödinger's equation just says that the classical kinetic energy plus the potential energy makes the total energy. This is basically a statement from classical mechanics, ultimately derivable from Newton's laws.

Observe that if V(r,t) is time invariant, and so reduces to just V(r), then |Ψ(r,t)| is also time invariant, and reduces to just |Ψ(r)|. That is, time then enters into the wave function Ψ(r,t) only through its un-observable phase factor. That is why Ψ(r,t) represents a *stationary* state for the Hydrogen atom.

Observe too that, however one writes Schrödinger's equation, there is no parameter cc, or any other trace of Maxwell theory in it. That is why Schrödinger's equation cannot give any clue about sub-states of the Hydrogen atom.

Observe next that, in giving only stationary Ψ(r,t) solutions, Schrödinger's equation does not reveal the irreversibility that we know exists in our macroscopic world. Lunin (2012) has identified this absence of irreversibility in Schrödinger's equation as an unsolved problem. Skála & Kapsa (2012) have noted that the measuring apparatus is not described in QM, and they have worked out an approach to deal with that deficit. Streklas (2012) has done the same with regard to the surrounding environment in which the system sits.

(An aside: the background could include a gravitational field, and Streklas notes how classical General Relativity Theory (GRT) breaks down at the Planck scale. As a cause for this failure, I suspect the SRT background from which GRT developed.)

In the present exposition, the time variation of V(r,t)V(r,t) can represent the intrusion of a measurement process, or an environmental factor, or some other disturbance, and then Schrödinger's equation can capture the phenomenon of irreversibility.

Can the foundation for Schrödinger's equation be further explicated by using the proposed Maxwell photon model? Recall that in its emission / propagation / reception scenario, the Maxwell photon model naturally displays particle-like localization at the two ends, and wave-like periodicity in the middle. The middle part of the photon scenario establishes a precedent for he use of the wave function as the subject of the Schrödinger equation.

The Maxwell photon model also helps explain why Schrödinger's equation seemingly demands complex numbers, while Physics before that time used them for convenience, but not out of necessity. Recall that that the Maxwell photon model has a second E, BE, B vector pair a quarter cycle out of phase with the first E, BE, B vector pair to make the circular polarization. That sort of phase issue naturally brings complex numbers.

Also, recall that the important output from the Maxwell photon model was its energy density, defined in terms of squared electric and magnetic fields. If we represented the fields a quarter cycle out of phase as imaginary numbers, then we would need fields, not just squared, but multiplied by complex conjugate fields. That operation would resemble the familiar |Ψ|2=ΨΨ*| operation for probability density.

Finally, the Maxwell photon model can help clarify the issue of 'duality'. The word has been taken to suggest a mysteriously simultaneous wave-particle character. But the general Schrödinger equation, with a time-dependent V(r,t) to represent some sort of measurement process, could certainly display the same less mysterious, more pedestrian, kind of duality that the Maxwell photon displays: sequential particle-like and wave-like behaviors.

Schrödinger's equation gives a lot more than just the ground state of Hydrogen. Like Maxwell's equations, it admits an infinite set of solutions. They are currently understood as representing an infinite set of excited states of the Hydrogen atom.

What exactly *are* excited states of the Hydrogen atom? The usual understanding is that they refer to something like spherical neighborhoods around the nucleus at larger radii, and that the electron can live in any one of these neighborhoods, and if it tumbles into a lower neighborhood, then a photon will be released.

I want to encourage readers to consider also any and all alternative interpretations that may be offered for the meaning of the term 'excited state'. My own working idea (Whitney 2012) is that 'excited state' does *not* refer to an attribute that a single Hydrogen atom can possess. The Hydrogen atom is too simple; it has too few degrees of freedom. My mental image of 'excited state' is a system involving, not one, but *several*, Hydrogen atoms.

The basis for such a candidate interpretation is that a balance between radiation and torquing works out, not only for two charges of opposite sign, but also for two, or more, charges of the *same* sign – if superluminal orbit speeds are allowed. And what is there to disallow them? The only factors are Einstein's Second Postulate and his resulting SRT, which together embed a rash denial of the well-known Arrow of Time. So be prudent; don't *a priori* disallow super-luminal orbit speeds.

The idea of the excited state of an atom as being actually a system of several atoms answers a need that was identified in Gevorkyan (2012). He pointed to spontaneous transition between quantum levels of a system as a hard-to-explain phenomenon. Indeed it *is* hard to explain in terms of excited states of a single atom. But it is easy to explain in terms of a system involving several atoms: the system can simply disintegrate back into several isolated single atoms.

5. QUANTUM CHEMISTRY

The main thing for Chemistry is not revealed in the Hydrogen atom, due to the fact that the Hydrogen atom has only one electron. The main thing for Chemistry is the variety of relationships among multiple electrons. Chemistry data suggest that electrons can sometimes actually attract, rather than repel, each other. That propensity can be important in driving chemical reactions. A basis for understanding that process is needed, and it cannot be found in the Schrödinger equation, or in any extension of Quantum Mechanics that injects Special Relativity Theory.

Buzea, Agop, & Nejneru (2012) investigate the Bohm/Vigier approach and the Madelung approach for the kinds of problems that Chemistry presents. The former approach relies at its outset on SRT, which seems risky to those of us who doubt SRT. The latter approach invokes a 'quantum potential' for interaction with a 'subquantic medium'. That sounds like 'aether', and seems risky to those of us who doubt the existence of 'aether'.

So perhaps additional approaches are still to be welcomed. One such approach was introduced inWhitney (2012), is expanded in Whitney (2013, in press), and is discussed further below.

The approach is called Algebraic Chemistry (AC). The name reflects the fact AC is carried out entirely with algebra, and not numerical integration. In fact, the math is hardly even algebra, since only the occasional square root goes beyond simple Arithmetic. Such simple math suffices because the AC approach is based on scaling laws. The model for the Hydrogen atom is the prototype for similar models of the atoms of all the other elements. The input information for all atoms is consists ionization potentials. The raw data set looks quite daunting, but as reported in Whitney (2012), the data fall into neat patterns when scaled by M/ZM/Z where MM is the nuclear mass number and ZZ is the nuclear charge.

This M/ZM/Z scaling produces a variable IP1,ZIP1,Z that we call 'population generic' because information about any element can be inferred from information about other elements. The AC Hydrogen-based model invites the division of each ionization potential IP1,ZIP1,Z into two parts, one part being IP1,1IP1,1 for the Hydrogen-like interaction of the electron population as a whole with the nucleus, and the other part being the increment ΔIP1,Z=ΔIP1,Z= IP1,Z−IP1,1IP1,Z−IP1,1 for the electron-electron interactions.

Modeling the energy requirements for making ions

The $IP_{1,Z}$ and $\Delta IP_{1,Z}$ for 118 elements are given in Whitney (2012). The utility of that data lies in the larger universe of inferences it supports. It can be used to estimate the actual energy involved in creating any ionization state of any element. To refer to the data $IP_{1,Z}$ and $\Delta IP_{1,Z}$ used, the term 'population generic' applies. To refer to the inferences made, the term 'element-specific' applies.

The formulae used are essentially the same for every element, so let us use the symbol 'EZ' for an arbitrary element, so we can write the formulae in a symbolic way. First consider the transition $_Z\mathrm{E} \rightarrow {}_Z\mathrm{E}^{+}$. It definitely takes an energy investment of $IP_{1,Z} \times Z / M_{Z'}$, where the factors of Z and $1/M_Z$ restore the population-generic information $IP_{1,Z}$ to element-specific information. This energy investment corresponds to a potential 'wall' to be gotten over. The wall has two parts, $IP_{1,1} \times Z / M_Z$ and $\Delta IP_{1,Z} \times Z / M_Z$. The transition $_Z\mathrm{E} \rightarrow {}_Z\mathrm{E}^{+}$ may also consume some heat, or generate some heat, as the remaining Z–1 electrons form new relationships, not necessarily instantaneously. This process constitutes adjustment to the rock pile, or the ditch, on the other side of the potential wall. It is represented by a term $-\Delta IP_{1,Z-1} \times (Z-1)/M_{Z'}$, where the factors of (Z–1) and 1/MZ restore the population-generic information $-\Delta IP_{1,Z-1}$ to element-specific information tailored for EZEZ. Thus altogether, $_Z\mathrm{E} \rightarrow {}_Z\mathrm{E}^{+}$ takes:

$$IP_{1,1} \times Z / M_Z + \Delta IP_{1,Z} \times Z / M_Z - \Delta IP_{1,Z-1} \times (Z-1)/M_Z.$$

Now consider removal of a second electron, $_Z\mathrm{E}^{+} \rightarrow {}_Z\mathrm{E}^{2+}$. Being already stripped of one of its electrons, the $_Z\mathrm{E} \rightarrow {}_Z\mathrm{E}^{+}$ system has less internal Coulomb attraction than neutral EZ has. So the factor of ZZ multiplying IP1,1 for $_Z\mathrm{E} \rightarrow {}_Z\mathrm{E}^{+}$ has to change to something smaller. Since Coulomb attraction generally reflects the product of the number of positive charges (here Z) and the number of negative charges (here Z–1), the reduced factor is $\sqrt{Z \times (Z-1)}$. Given this factor, $_Z\mathrm{E}^{+} \rightarrow {}_Z\mathrm{E}^{2+}$ takes:

$$IP_{1,1} \times \sqrt{Z \times (Z-1)}/M_Z + \Delta IP_{1,Z-1} \times (Z-1)/M_Z - \Delta IP_{1,Z-2} \times (Z-2)/M_Z$$

Observe that putting the steps $_ZE \rightarrow {}_ZE^{+}$ and $_ZE^{+} \rightarrow {}_ZE^{2+}$ together, the terms involving $\Delta IP_{1,Z-1}$ cancel, leaving that altogether, $_ZE^{+} \rightarrow {}_ZE^{2+}$ takes:

$$IP_{1,1} \times Z/M_Z + IP_{1,1}\sqrt{Z \times (Z-1)}/M_Z + \Delta IP_{1,Z} \times Z/M_Z - \Delta IP_{1,Z-2} \times (Z-2)/M_Z.$$

This reduction to just two terms involving ΔIPs is typical of all sequential ionizations, of however many steps. Observe too that in the cumulative, we have *two* terms in IP1,1 $IP_{1,1} \times Z/M_Z$ and $IP_{1,1}\sqrt{Z \times (Z-1)}/M_Z$. This makes sense since two electrons are removed.

From the above, it should be clear how to proceed with stripping however many more electrons you may be interested in removing. Now consider adding an electron to EZ. The problem is similar to removing an electron from $_{Z+1}E$,1, but in reverse. So, $_ZE^{-} \rightarrow {}_ZE^{2-}$ takes:

$$-IP_{1,1} \times \sqrt{Z \times (Z+2)}/M_Z - \Delta IP_{1,Z+2} \times (Z+2)/M_Z + \Delta IP_{1,Z+1} \times (Z+1)/M_Z$$

Going one step further, the problem of adding another electron to EZ− is similar to removing an electron from $_{Z+2}E$, but in reverse. So, $_ZE^{-} \rightarrow {}_ZE^{2-}$ takes:

$$-IP_{1,1} \times \sqrt{Z \times (Z+1)}/M_Z - IP_{1,1} \times \sqrt{Z \times (Z+2)}/M_Z + \Delta IP_{1,Z} \times Z/M_Z - \Delta IP_{1,Z+2} \times (Z+2)/M_Z$$ O

bserve that putting the steps $_ZE \rightarrow {}_ZE^{-}$ and $_ZE^{-} \rightarrow {}_ZE^{2-}$ together, the terms involving $\Delta IP_{1,Z+1}$ cancel, leaving that altogether, $_ZE^{-} \rightarrow {}_ZE^{2-}$ takes:

$$-IP_{1,1} \times \sqrt{Z \times (Z+1)}/M_Z - IP_{1,1} \times \sqrt{Z \times (Z+2)}/M_Z + \Delta IP_{1,Z} \times Z/M_Z - \Delta IP_{1,Z+2} \times (Z+2)/M_Z$$

From the above, it should be clear how to proceed with adding however many more electrons you may be interested in adding.

To illustrate the development and use of information, consider a few example elements: Hydrogen, Carbon and Oxygen. The steps to develop essential information for Hydrogen are:

Write Formulae:

$_1H \to {_1H^+}{:}IP_{1,1}\times 1/M_1+\Delta IP_{1,1}\times 1/M_1-\Delta IP_{1,0}\times 0/M_1$(Note: $\Delta IP_{1,1}\equiv 0$, and $\Delta IP_{1,0}$ does not exist.)

$_1H \to {_1H^-}{:}-IP_{1,1}\times\sqrt{1\times 2}/M_1-\Delta IP_{1,2}\times 2/M_1+\Delta IP_{1,1}\times 1/M_1$

Insert Data:

$_1H \to {_1H^+}{:}14.250\times 1/1.008+0-0$

$_1H \to {_1H^-}{:}-14.250\times 1.4142/1.008-35.625\times 2/1.008+0\times 1/1.008$

Evaluate Formulae:

$_1H \to {_1H^+}{:}14.1369+0-0=14.1369\text{eV}$

$_1H \to {_1H^-}{:}-19.9924-70.6845+0=-90.6769\text{eV}$

The steps to develop the information needed for Carbon are far more numerous because it routinely gives or takes so many electrons. The steps are:

Write Formulae:

$_6C \to {_6C^+}{:}IP_{1,1}\times 6/M_6+\Delta IP_{1,6}\times 6/M_6-\Delta IP_{1,5}\times 5/M_6$

$_6C^+ \to {_6C^{2+}}{:}IP_{1,1}\times\sqrt{6\times 5}/M_6+\Delta IP_{1,5}\times 5/M_6-\Delta IP_{1,4}\times 4/M_6$

$_6C^{2+} \to {_6C^{3+}}{:}IP_{1,1}\times\sqrt{6\times 4}/M_6+\Delta IP_{1,4}\times 4/M_6-\Delta IP_{1,3}\times 3/M_6$

$$_6C^{3+} \rightarrow {}_6C^{4+}: IP_{1,1} \times \sqrt{6\times 3} / M_6 + \Delta IP_{1,3} \times 3 / M_6 - \Delta IP_{1,2} \times 2 / M_{16}$$

$$_6C \rightarrow {}_6C^{-}: -IP_{1,1} \times \sqrt{6\times 7} / M_6 - \Delta IP_{1,7} \times 7 / M_6 + \Delta IP_{1,6} \times 6 / M_6$$

$$_6C^{-} \rightarrow {}_6C^{2-}: -IP_{1,1} \times \sqrt{6\times 8} / M_6 - \Delta IP_{1,8} \times 8 / M_6 + \Delta IP_{1,7} \times 7 / M_6$$

$$_6C^{2-} \rightarrow {}_6C^{3-}: -IP_{1,1} \times \sqrt{6\times 9} / M_6 - \Delta IP_{1,9} \times 9 / M_6 + \Delta IP_{1,8} \times 8 / M_6$$

$$_6C^{3-} \rightarrow {}_6C^{4-}: -IP_{1,1} \times \sqrt{6\times 10} / M_6 - \Delta IP_{1,10} \times 10 / M_6 + \Delta IP_{1,9} \times 9 / M_6$$

Insert Data:

$$_6C \rightarrow {}_6C^{+}: 14.250 \times 6 / 12.011 + 7.320 \times 6 / 12.011 - 2.805 \times 5 / 12.011$$

$$_6C^{+} \rightarrow {}_6C^{2+}: 14.250 \times 5.4772 / 12.011 + 2.805 \times 5 / 12.011 - 9.077 \times 4 / 12.011$$

$$_6C^{2+} \rightarrow {}_6C^{3+}: 14.250 \times 4.8990 / 12.011 + 9.077 \times 4 / 12.011 - (-1.781) \times 3 / 12.011$$

$$_6C^{3+} \rightarrow {}_6C^{4+}: 14.250 \times 4.2426 / 12.011 + (-1.781) \times 3 / 12.011 - 35.625 \times 2 / 12.011$$

$$_6C \rightarrow {}_6C^{-}: -14.250 \times 6.4807 / 12.011 - 13.031 \times 7 / 12.011 + 7.320 \times 6 / 12.011$$

$$_6C^{-} \rightarrow {}_6C^{2-}: -14.250 \times 6.9282 / 12.011 - 13.031 \times 8 / 12.011 + 13.031 \times 7 / 12.011$$

$$_6C^{2-} \rightarrow {}_6C^{3-}: -14.250 \times 7.3485 / 12.011 - 20.254 \times 9 / 12.011 + 13.031 \times 8 / 12.011$$

$$_6C^{3-} \rightarrow {}_6C^{4-}: -14.250 \times 7.7460 / 12.011 - 29.391 \times 10 / 12.011 + 20.254 \times 9 / 12.011$$

Evaluate Formulae:

$$_6C \rightarrow {}_6C^{+}: 7.1185 + 3.6566 - 1.1677 = 9.6074 eV$$

$$_6C^{+} \rightarrow {}_6C^{2+}: 6.4982 + 1.1678 - 3.0229 = 4.6431 eV$$

$$_6C^{2+} \rightarrow {}_6C^{3+}: 5.8122 + 3.0229 + 0.4448 = 9.2799 eV$$

$$_6C^{3+} \rightarrow {}_6C^{4+}: 5.0335 - 0.4448 - 5.9321 = -1.3434 eV$$

$$_6C \rightarrow {}_6C^{-}: -7.6888 - 7.5945 + 3.6566 = -11.6267 eV$$

$$_6C^{-} \rightarrow {}_6C^{2-}: -8.2197 - 8.6794 + 7.5945 = -9.3046 eV$$

$$_6C^{2-} \rightarrow {}_6C^{3-}: -8.7184 - 15.1766 + 8.6794 = -15.2156 eV$$

$$_6C^{3-} \rightarrow {}_6C^{4-}: -9.1900 - 24.4701 + 15.1766 = -18.4835 eV$$

Evaluate sums:

$${}_6C \rightarrow {}_6C^{2+}: 9.6074 + 4.6431 = 14.2505\text{eV}$$

$${}_6C \rightarrow {}_6C^{3+}: 14.2505 + 9.2799 = 23.5304\text{eV}$$

$${}_6C \rightarrow {}_6C^{4+}: 23.5304 - 1.3434 = 22.187\text{eV}$$

$${}_6C \rightarrow {}_6C^{2-}: -11.6267 - 9.3046 = -20.9313\text{eV}$$

$${}_6C \rightarrow {}_6C^{3-}: -20.9313 - 15.2156 = -36.1469\text{eV}$$

$${}_6C \rightarrow {}_6C^{4-}: -36.1469 - 18.4835 = -54.6304\text{eV}$$

The steps to develop information for Oxygen are:

Write Formulae:

$${}_8O \rightarrow {}_8O^{+}: IP_{1,1} \times 8 / M_8 + \Delta IP_{1,8} \times 8 / M_8 - \Delta IP_{1,7} \times 7 / M_8$$

$${}_8O^{+} \rightarrow {}_8O^{2+}: IP_{1,1} \times \sqrt{8 \times 7} / M_8 + \Delta IP_{1,7} \times 7 / M_8 - \Delta IP_{1,6} \times 6 / M_8$$

$${}_8O \rightarrow {}_8O^{-}: -IP_{1,1} \times \sqrt{8 \times 9} / M_8 - \Delta IP_{1,9} \times 9 / M_8 + \Delta IP_{1,8} \times 8 / M_8$$

$${}_8O^{-} \rightarrow {}_8O^{2-}: -IP_{1,1} \times \sqrt{8 \times 10} / M_8 - \Delta IP_{1,10} \times 10 / M_8 + \Delta IP_{1,9} \times 9 / M_8$$

Insert Data:

$${}_8O \rightarrow {}_8O^{+}: 14.250 \times 8 / 15.999 + 13.031 \times 8 / 15.999 - 13.031 \times 7 / 15.999$$

$${}_8O^{+} \rightarrow {}_8O^{2+}: 14.250 \times 7.4833 / 15.999 + 13.031 \times 7 / 15.999 - 7.320 \times 6 / 15.999$$

$${}_8O \rightarrow {}_8O^{-}: -14.250 \times 8.4853 / 15.999 - 20.254 \times 9 / 15.999 + 13.031 \times 8 / 15.999$$

$${}_8O^{-} \rightarrow {}_8O^{2-}: -14.250 \times 8.9443 / 15.999 - 29.391 \times 10 / 15.999 + 20.254 \times 9 / 15.999$$

Evaluate Formulae:

$${}_8O \rightarrow {}_8O^{+}: 7.1254 + 6.5159 - 5.7014 = 7.9399 eV$$

$${}_8O^{+} \rightarrow {}_8O^{2+}: 6.6652 + 5.7014 - 2.7452 = 9.6214 eV$$

$${}_8O \rightarrow {}_8O^{-}: -7.5577 - 11.3936 + 6.5159 = -12.4354 eV$$

$${}_8O^{-} \rightarrow {}_8O^{2-}: -7.9665 - 18.3705 + 11.3936 = -14.9434 eV$$

Evaluate sums:

$${}_8O \rightarrow {}_8O^{2+}: 7.9399 + 9.6214 = 17.5613 eV$$

$${}_8O \rightarrow {}_8O^{2-}: -12.4354 - 14.9434 = -27.3788 eV$$

What can we tell from all this information? Consider a few of the molecules that these elements can make. The simplest one is the Hydrogen molecule H2H2. Forming it takes 14.1369−90.6769= −76.5414.1369−90.6769=−76.54eV. That number is very negative, which means the Hydrogen molecule forms quickly, even explosively. Isolated neutral Hydrogen atoms are rare in Nature. Even at very low density, in deep space, Hydrogen atoms would rather form molecules, or form plasma, than remain as neutral atoms. How ironic it is that the prototypical atom for the development of QM was something that is hardly to be found in Nature!

Another simple molecule is O2O2. This oxygen molecule illustrates the interesting possibility of more than one ionic configuration, a situation that turns out to be the case for many molecules. O2 can be ${}_8O^{+} + {}_8O^{-}$ or ${}_8O^{2+} + {}_8O^{2-}$. Forming ${}_8O^{+} + {}_8O^{-}$ takes 7.9399−12.4354= −4. 49557.9399−12.4354=−4.4955eV, and forming ${}_8O^{+} + {}_8O^{-}$ or ${}_8O^{2+} + {}_8O^{2-}$. takes 17.5613−27.3788=−9.817517.5613− 27.3788= −9.8175eV. Although the second ionic configuration is better in terms of energy, the two are close. In situations like this, both ionic configurations exist, in proportions determined by thermodynamic entropy maximization, which depends on temperature.

The fact that the two ionic configurations of O2 are so close in energy means that transitions between them are easy. This can make O2 an absorber and emitter of low energy photons; *i.e.* infrared photons, *i.e.* heat. This can in turn make Oxygen act as a so-called 'greenhouse gas'; *i.e.*, a contributor to atmospheric warming. But as O2O2-consuming animals, we just never speak of O2O2 in such a derogatory way.

Consider H2OH2O. Water again illustrates the possibility of more than one ionic configuration for a given molecule. Water is known to dissociate into the naked proton H+H+ and the hydroxyl radical OH−OH−. In turn, the hydroxyl radical has to be the combination of ions $2H^{-}+O^{2+}$; there is not an alternative form using H−H−, because then OO would have to be neutral. So we might well imagine that H2O had ionic configuration $2H^{+}+O^{2-}$. But the formation of that ionic configuration takes 2×14.1369−27.3788=+0.89502×14.1369−27.3788=+0.8950eV, a slightly positive energy. That can't be right for the common water molecule covering our planet. So in fact, common water must not live in the ionic configuration to which it dissociates, and thereby dies. Therefore, consider the alternative ionic configuration $2H^{-}+O^{2+}$ .This one requires 2×(−90.6769)+17.5613=2×(−90.6769)+17.5613=−163.7925−163.7925eV. This is a decidedly negative energy, and so is believable for a decidedly stable molecule.

Water in the normal $2H^{-}+O^{2+}$ ionic configuration has to form a tetrahedron, with two H+H+ naked protons on two vertices and two 2e−2e− electron pairs on and the other two vertices. That is why the water molecule we know has a bend to it. Viewing the Hydrogen nuclei as lying on arms originating from the Oxygen nucleus, the angle of the bend is the angle characteristic of arms from the center to two corners of a regular tetrahedron – on the order of 109.5°109.5°.

The other ionic configuration for water, $2H^{+}+O^{2-}$ is the charge mirror image of the commonly known one. But it has to be a completely different shape: not tetrahedral, but instead linear. It just looks like H · O · HH · O · H, where the dots mean 'chemical bond'. This form of water apparently does exist, but only in a very un-natural circumstance. There exists an electrochemically created substance known as 'Brown's gas' that has occasioned some impossibly wild claims about energy generation, but has also been investigated quite legitimately for applications in welding. A linear isomer of water is thought to be the active ingredient in Brown's gas.

The story of water tells us that even the most familiar of compounds can have some very interesting isomers. The conclusion to be drawn is that *any* molecule with three or more atoms can have isomers that differ, certainly in ionic configuration, but probably also in molecular shape, and in resultant chemical properties.

The very negative −163.7925−163.7925eV result for forming the $2H^{-}+O^{2+}$ionic configuration of normal water is what makes the burning of all hydrocarbons so worthwhile as energy sources. It is the main thing, but it generally attracts no mention. Other reaction products get all the attention.

Consider CO2CO2. Carbon dioxide is a normal atmospheric constituent, and a product of hydrocarbon combustion, a possible contributor to global warming, and sometimes a target for government regulation.

CO_2 again illustrates the possibility of several ionic configurations for a molecule. CO2CO2 has at least four plausible ionization configurations: $C^{4+}+2O^{2-}$, $C^{4-}+2O^{2+}$, $C^{2+}+2O^{-}$, and $C^{2-}+O^{+}$., and $C^{2-}+O^{+}$The energy requirements to make them from the neutral atoms are:

$$C^{4+}+2O^{2-}: 22.187+2\times(-27.3788)=22.187-54.7576=-32.5706\text{eV}$$

$$C^{4-}+2O^{2+}: -54.6304+2\times17.5613=-54.6304+35.1226=-19.5078\text{eV}$$

$$C^{2+}+2O^{-}: 14.2505+2\times(-12.4354)=14.2505-24.8708=-10.6203\text{eV}$$

$$C^{2-}+2O^{+}: -20.9313+2\times7.9399=-20.9313+15.8798=-5.0515\text{eV}$$

The first ionic configuration listed is the one favored electrically, but the others must also occur, all in thermodynamically determined proportion. The fact that there are so *many* possibilities for just this one little tri-atomic molecule means that QC can benefit from using AC to identify all the possibilities in a situation, and select rationally among them, and spend computation power wisely.

Single-electron state filling over the periodic table

Nobody is yet satisfied that we completely understand the Periodic Table. QM informs us of single electron states and their quantum numbers, and we can tell from spectroscopic data what single electron states are filled for each element, and we can see what the governing rule is, but we do not understand why that is the rule, and we do not understand why there are exceptions to the rule.

The normal order of state filling can be described in terms of the quantum numbers n=1,2,3,... for radial level, l=0,1,...n−1 for orbital angular momentum, and s=−1/2,+1/2 for spin. The normal order of state filling goes with increasing n+ln+l, with all of s=−1/2 first, and then all of s=−1/2. So that makes the normal order:

In Period 1: $n=1,\ \ l=0,\ \ s=-1/2$ and $n=1,\ \ l=0,\ \ s=+1/2$; Then in Period 2:

$n=2,\ \ l=0,\ \ s=-1/2,\ \ \ n=2,\ \ l=0,\ \ s=+1/2$ and $n=2,\ \ l=1,\ \ s=-1/2$ three times, $n=2,\ \ l=1,\ \ s=+1/2$ three times;

Then in Period 3:

$n=3,\ \ l=0,\ \ s=-1/2,\ \ \ n=3,\ \ l=0,\ \ s=+1/2$ and $n=3,\ \ l=1,\ \ s=-1/2$ three times, $n=3,\ \ l=1,\ \ s=+1/2$ three times;

Then in Period 4:

$n=4,\ \ l=0,\ \ s=\pm1/2,\ n=3,\ \ l=2,\ \ s=-1/2$ five times, $n=3,\ \ l=2,\ \ s=+1/2$ five times,

$n=4,\ \ l=1,\ \ s=-1/2$ three times, $n=4,\ \ l=1,\ \ s=+1/2$ three times; ;

Then in Period 5:

$n=5,\ \ l=0,\ \ s=\pm1/2,\ n=4,\ \ l=2,\ \ s=-1/2$ five times, $n=4,\ \ l=2,\ \ s=+1/2$ five times,

$n=5,\ \ l=1,\ \ s=-1/2$ three times, $n=5,\ \ l=1,\ \ s=+1/2$ three times;

Then in Period 6:

$n=6,\ \ l=0,\ \ s=\pm1/2,\ n=5,\ \ l=3,\ \ s=-1/2$ seven times, $n=5,\ \ l=3,\ \ s=+1/2$ seven times,

$n=5,\ \ l=3,\ \ s=-1/2$ five times, $n=5,\ \ l=3,\ \ s=+1/2$ five times,

$n=6,\ \ l=1,\ \ s=-1/2$ three times, $n=6,\ \ l=1,\ \ s=+1/2$ three times;

Then in Period 7:

$n=7,\ l=0,\ s=\pm 1/2, n=6,\ l=3,\ s=-1/2$ seven times, $n=6,\ l=3,\ s=+1/2$ seven times,

$n=6\ l=3,\ s=-1/2$ five times, $n=6\ l=3,\ s=+1/2$ five times,

$n=7,\ l=1,\ s=-1/2$ three times, $n=7,\ l=1,\ s=+1/2$ three times.

It is perhaps possible to do enough QM calculations to develop a numerical explanation for this pattern. But we do not have from QM any higher-level, conceptual explanation for this pattern.

And on top of that, there are 19 exceptions to the pattern. They are:

In Period 4: Chromium $_{24}$Cr, Copper $_{29}$Cu; ;

In Period 5: Niobium $_{41}$bN , Molybdenum $_{42}$Mo, Rubidium $_{44}$Ru, Rhodium 45 ,

In Period 6: Lanthanum 57LL , Cerium $_{58}$Ce , Gadolinium $_{64}$Gd, Platinum $_{78}$Pt, Gold $_{79}$Au ;

In Period 7: Actinium $_{89}$Ac, Thallium $_{90}$Th, Protactinium $_{91}$Pa, Uranium $_{92}$UU, Neptunium $_{93}$Np, Cerium $_{96}$Cm.

So it is a good project for AC to try to improve this situation, both in regard to explaining the pattern, and in regard to explaining the exceptions.

The Hydrogen-based model used for AC makes the electron population a rather localized subsystem, orbiting the nucleus, rather than enclosing it. The electron subsystem is composed of electron rings spinning at superluminal speeds, stacked together like little magnets. This model creates a hierarchy of magnetic confinement levels. Two rings with two electrons each create a 'magnetic bottle', and it can contain up to two geometrically smaller rings with three electrons each. Two such three-electron rings create a stronger 'magnetic thermos jug'. That can contain up to two geometrically smaller rings with five electrons each. Two such five-electron rings create an even stronger 'magnetic Dewar flask'. It is capable of containing up to two geometrically smaller rings with seven electrons each.

The electron state filling sequence is determined by a rather 'fractal' looking algorithm: Always build and store an electron ring for the largest number of electrons possible, where 'possible' means having a suitable magnetic confinement volume available to fit into, and where 'suitable' means created by two electron rings with smaller electron count, and not yet filled with two electron rings of larger electron count. Sometimes only a new two-electron ring is possible, and that is what starts a new period in the Periodic Table.

So there follows the expected order for the filling of single electron states across the Periodic Table.

AC can also identify factors that account for individual exceptions. The worst exception is Palladium, because it has not just one, but two, violations of the nominal pattern. Here is the explanation for Palladium. According to the nominal electron filling pattern, P46dP46d would have a not yet used space for two three-electron rings, and hence it would have an un-used two-electron ring. Total consumption of that un-used two-electron ring into an unfilled five-electron ring allows an extremely symmetric stack of filled electron rings. It goes: 2, 3, 5,5, 3, 2, 3,3, 2, 3, 5,5, 3, 2. The opportunity for such symmetry is what trumps the nominal pattern.

Here is a list of brief comments about all of the exceptions:

Chromium $_{24}$Cr robs one electron from a two-electron ring to complete a five-electron ring.

Copper $_{29}$Cu robs one electron from a two-electron ring to complete a five-electron ring.

Niobium $_{41}$ Nb robs one electron from a two-electron ring to complete a five-electron ring.

Molybdenum $_{42}$Mo robs one electron from a two-electron ring to complete a five-electron ring.

Rubidium R$_{44}$u robs one electron from a two-electron ring to complete a five-electron ring.

Rhodium $R_{45}hR45h$ robs one electron from a two-electron ring to complete a five-electron ring.

Palladium $P_{46}d$ completely consumes a two-electron ring to complete a five-electron ring.

Silver $A_{47}g$ robs one electron from a two-electron ring to complete a five-electron ring.

Lanthanum $L_{57}a$ puts an electron in a five-electron place instead of a seven-electron place.

Cerium $C_{58}eC58e$ puts an electron in a five-electron place instead of a seven-electron place.

Gadolinium $G_{64}d$ puts an electron in a five-electron place instead of a seven-electron place.

Platinum $P_{78}tP$ robs one electron from a two-electron ring to complete a five-electron ring.

Gold $A_{79}u$ robs one electron from a two-electron ring to complete a five-electron ring.

Actinium $A_{89}c$ puts an electron in a five-electron place instead of a seven-electron place.

Thallium $T_{90}hT90h$ puts an electron in a five-electron place instead of a seven-electron place.

Protactinium $_{91}$ Pa puts an electron in a five-electron place instead of a seven-electron place.

Uranium $_{92}$ UU puts an electron in a five-electron place instead of a seven-electron place.

Neptunium $_{93}$ NpN93p puts an electron in a five-electron place instead of a seven-electron place.

Cerium $_{96}$Cm puts an electron in a five-electron place instead of a seven-electron place.

6. CONCLUSION

It is this author's opinion is that a more fully developed quantum mechanics, giving equal attention to both stable states, *and* the transitions between them, would fulfill a property that the subject matter of quantum mechanics has always demanded: some kind of duality. All of the objects of study in quantum mechanics exhibit a wave-particle duality, and the theory itself needs a corresponding kind of duality: attention both to the definition of stable states, and to the study of details of state transitions.

This paper attempts to make some progress in that direction. It gives several examples of old problems treated with a new approach. The first of these concerns the nature of the photon. There is a perception that the discovery of the photon marks a departure from Maxwell. This author disagrees with that perception. The second problem concerns Schrödinger's equation. There is a perception that Schrödinger's equation marks a departure from Newton, and the classical physics of particles. This author disagrees with that perception too. The third problem has to do with the application of QM in Chemistry. There is a perception that such an application of QM demands extensive computer calculation. This author believes in an alternative approach based on scaling laws: Algebraic Chemistry.

All of these problems illustrate a reasoned concern about the current practice of Physics. I think Physics sometimes goes a step too far in the direction of reductionism. Einstein's signal for SRT, with only a speed parameter, is a step too far. The photon concept without Maxwell's equations is a step too far. An atom with a stationary nucleus is a step too far. Schrödinger's equation without the $\Psi^*\Psi^*$ factor is a step too far. A single atom with multiple excited states is a step too far. And so on.

Consider some history. At the turn of the 20th century, there was a lot of work concerning a isolated electron. Why does the electron not explode due to internal Coulomb forces? Why does the equation of motion for the electron allow run-away solutions that do not occur in Nature? There are many such puzzles. Chapter 17 in the standard textbook by Jackson (1975) discusses radiation damping, self-fields of a particle, scattering and absorption of radiation by a bound system, all from the classical viewpoint and from the viewpoint of SRT. It is a status report, not a final resolution. These matters are still not fully resolved. I suspect too much reductionism as their cause.

Always remember: Sometimes, backing off from reductionism, and analyzing a slightly more complicated problem, actually leads to simpler results.

REFERENCES

1. Abad, L.V., & S. C. Huichalaf (2012) Complementarity in Quantum Mechanics and Classical Statistical Mechanics, Chapter 1 in *Theoretical Concepts of Quantum Mechanics*, Ed. M.R. Pahlavani, InTech.

2. Berkdemir, C. (2012) Application of the Nikiforov-Uvarov Method in Quantum Mechanics, Chapter 11 in *Theoretical Concepts of Quantum Mechanics*, Ed. M.R. Pahlavani, InTech.

3. Bracken, P. (2012) Quantum Mechanics Entropy and a Quantum Version of the H theorem, Chapter 20 in *Theoretical Concepts of Quantum Mechanics*, Ed. M.R. Pahlavani, InTech.

4. Buzea, Calin Gh., M. Agop, & C. Nejneru (2012) Correspondences of Scale Relativity Theory with Quantum Mechanics, Chapter 18 in *Theoretical Concepts of Quantum Mechanics*, Ed. M.R. Pahlavani, InTech.

5. Cini, M. (2012) The Physical Nature of Wave/Particle Duality, Chapter 2 in *Theoretical Concepts of Quantum Mechanics*, Ed. M.R. Pahlavani, InTech.

6. De Zela, F. (2012) The Pancharatnam-Berry Phase: Theoretical and Eperimental Aspects, Chapter 14 in *Theoretical Concepts of Quantum Mechanics*, Ed. M.R. Pahlavani, InTech.

7. Gevorkyan, A.S. (2012) Nonrelativistic Quantum Mechanics with Fundamental Environment, Chapter 8 in *Theoretical Concepts of Quantum Mechanics*, Ed. M.R. Pahlavani, InTech.

8. Jackson, J.D. (1975) *Classical Electrodynamics*, John Wiley & Sons, New York, *etc.*

9. Lunin, N.V. (2012) The Group Theory and Non-Euclidean Superposition Principle in Quantum Mechanics, Chapter 13 in *Theoretical Concepts of Quantum Mechanics*, Ed. M.R. Pahlavani, InTech.

10. Skála, L., & V. Kapsa (2012) Quantum Mechanics and Statistical Description of Results of Measurement, Chapter 10 in *Theoretical Concepts of Quantum Mechanics*, Ed. M.R. Pahlavani, InTech.

11. Streklas, A. (2012) Non Commutative Quantum Mechanics in Time-Dependent Backgrounds, Chapter 9 in *Theoretical Concepts of Quantum Mechanics*, Ed. M.R. Pahlavani, InTech.

12. Torres-Vega, G. (2012), Correspondence, Time, Energy, Uncertainty, Tunnelling, and Collapse of Probability Densities, Chapter 4 in *Theoretical Concepts of Quantum Mechanics*, Ed. M.R. Pahlavani, InTech.

13. Whitney, C. K. (2012) Better Unification for Physics in General Through Quantum Mechanics in Particular, Chapter 7 in *Theoretical Concepts of Quantum Mechanics*, Ed. M.R. Pahlavani, InTech.

14. Whitney, C.K. (2013) *Algebraic Chemistry*: *Applications and Origins*, Nova Science Publishers.

INDEX

N

P

Q

S

T

U